THE ZOMBIE CURSE

THE ZOMBIE CURSE

A Doctor's 25-Year Journey into the
Heart of the AIDS Epidemic in Haiti

Arthur M. Fournier, M.D.

with

Daniel Herlihy

Joseph Henry Press
Washington, D.C.

Joseph Henry Press • 500 Fifth Street, NW • Washington, DC 20001

The Joseph Henry Press, an imprint of the National Academies Press, was created with the goal of making books on science, technology, and health more widely available to professionals and the public. Joseph Henry was one of the founders of the National Academy of Sciences and a leader in early American science.

Any opinions, findings, conclusions, or recommendations expressed in this volume are those of the author and do not necessarily reflect the views of the National Academy of Sciences or its affiliated institutions.

Library of Congress Cataloging-in-Publication Data

Fournier, Arthur M. (Arthur Michael), 1947-
  The zombie curse : a doctor's 25-year journey into the heart of the AIDS epidemic in Haiti / Arthur M. Fournier, with Daniel Herlihy.
    p. cm.
  Includes bibliographical references and index.
  ISBN 0-309-09736-3 (cloth : alk. paper)
  1. AIDS (Disease)—Haiti.   I. Herlihy, Daniel. II. Title.
  [DNLM: 1. Acquired Immunodeficiency Syndrome—epidemiology —Personal narratives. 2. Physicians. 3. Florida. 4. Haiti—epidemi-ology.   WC 503.4 DH2 F778z 2006]
  RA643.86.H35F68 2006
  362.196'97920097294—dc22
                                        2005031643

Cover design by Michele de la Menardiere. Photo ©Robert W. Kelley, Time and Life Pictures Collection.

Creole orthography by Jan Mapou.

Printed in the United States of America

Contents

Acknowledgments

I WOULD LIKE TO ACKNOWLEDGE Dan Herlihy, my best friend from college, for assisting me by critiquing, editing, and word-crafting this manuscript. Dan was an English major in college and taught English for 35 years. It's not easy being a first-time author. Without the encouragement he provided and the discipline he inflicted, this work would probably be unreadable, and assuredly a lot less interesting. In addition, Denise Fluellen-Bradley, Marie Bennett, and Sonia Page provided invaluable assistance in the preparation of the manuscript.

I would also like to acknowledge the Open Society Institute, Partners in Health, and the Green Family Foundation for their support of Medishare (my charity in Haiti). These are three very different organizations in terms of endowment, mission, and focus. Each empowered Medishare in a different way—OSI by funding our family medicine residency training program for Haitian doctors, Partners in Health for showing the way to develop a preferential option for the poor, and the Green Family Foundation by developing our community health program in Thomonde. With their help (and yours, since the royalties from this book will further support our work) the shaky little charity founded by Dr. Barth Green and myself 11 years ago can now stand on its own two feet. Finally, I would like to acknowledge the hours of dedicated service contributed to Medishare by doctors, nurses, medical students, and health care workers—Haitians and Americans—in partnership and solidarity with the most remarkable people in the Western Hemisphere.

Bon sante a tout moun!

Foreword

By Paul Farmer

EACH YEAR, THOUSANDS OF Americans come to Haiti to "help out." Our neighbor is, as the cliché goes, the poorest country in the hemisphere and one of the poorest in the world; surely it could use some help from its rich neighbor to the north. But what kind of help, exactly? My friend and colleague Arthur Fournier's answers to this question change subtly as the two decades chronicled in his book unfold. The answers change as he learns more about Haitians; of course Haiti and the United States, and their tortured relations, change, too. But *The Zombie Curse* is not a sociological or historical text, nor is it a report about Fournier's projects in Haiti. Fournier's book is a memoir, and, as in all honest memoirs, Fournier is really writing of his own personal experience as an American doctor living in a metropolis that is anything but culturally homogeneous. Miami is, as much as Haiti, at the center of this tale.

Fournier's is an often-riveting book, in great part because he tells us how he learned about AIDS from his patients—first in Miami and then in Haiti. The AIDS epidemic also changes with time, and the more time Fournier spends in Haiti, the more he senses that AIDS is, now more than ever, about poverty and inequality. This is a hard lesson for physicians to learn, and Fournier relates with admirable candor his own role in spreading, unintentionally, early inaccuracies about AIDS in Haiti. Fournier tells how he found himself quite unpopular among certain new acquaintances in Haiti, which had been inaccurately branded as the source of the much larger epidemic to the north. For a while, in Miami as elsewhere in the eastern United States, the Haiti-as-source-of-AIDS myth ruled supreme. Sitting with a group of Haitian professionals complaining about their stigmatization as a nation, Fournier "confessed my own role in the Haitian AIDS study." *How could you have done that?* comes their reply. It is this sort of candid detail that makes this book worth reading.

Fournier's first visits to Haiti are brief ones, as are those of most American do-gooders—I use this term without irony, for reasons to be revealed below. We hear, often and in this book, that such short trips change the lives of those, especially students, who take them. In my experience as a medical school professor, this is almost invariably true. But the trick is to change lives in Haiti, and to change them for the better. That's the hard part, and the discovery that changing lives is difficult is a leitmotiv of this memoir. Once in Haiti, Fournier learns about its cultural complexities—the ones not readily visible from Miami—and vertiginous class structure. He comes to see how violence is inevitably born of great inequality. This is another painful lesson, since his colleagues are largely from the professional classes; his patients, by and large, are the destitute. Violence is a recurring theme in the latter half of the book because Fournier's efforts, like those of most "humanitarian" groups—including ours—working in Haiti, have not yet been effective in diminishing the social divide that so damages that country. Fournier worries about the outbreak of "civil war," but in fact civil strife of this sort is unlikely in Haiti, since only one side is really armed. The violence comes mostly from above.

In this book, Fournier presents his work, and the work of our organization, Partners In Health, as successful; and of course this is gratifying to read. But humanitarian medicine alone will not suffice to stanch the wounds he sees everywhere in Haiti. We read that this term—"humanitarian medicine"—causes some academic conflict in Fournier's home institution as he and his co-workers seek to link their efforts to Miami's flagship university.

These debates were new to Fournier, he allows, but they are not new. David Rieff is mentioned in the book, and Rieff's cautionary work on humanitarian aid is well worth reading. But for the reader who wants to hear only about medicine and public health in settings marred by violence, listen to Dr. Gino Strada, an Italian surgeon who decided, after becoming a successful transplant surgeon, that he would try to help civilians—most of them children—injured by land mines and other man-made pathogens. After a decade of this grueling work, in Rwanda and Pakistan and many other places, what Strada found most

> worrying was the fact that many organizations, to which I find it hard to apply the adjective "humanitarian," had accepted money from the governments engaged in the war without a moment's hesitation. *Pecunia non olet!* (Money doesn't stink!) This was the philosophy adopted by a great many non-governmental organizations (NGOs), in a great hurry not to miss the opportunity, without pausing to reflect on their actions. And without detecting the trap.[1]

Such traps are set across Haiti, which probably boasts, per capita, more NGOs than any other poor country. And here's a dilemma for all of us: if there are so many NGOs, why is the situation still so grim? How might us do-gooders do better? Universities, too, are waking up to global health and to human rights work, and finding an academic field littered with landmines. Will the traditional products of a university—more studies, more seminars, more publications—be enough? The most experienced hands in poverty reduction are not always sanguine on this score. When Stephen Lewis, U.N. Special Envoy for HIV/AIDS in Africa, writes about the need for

---

[1]Strada G. *Green Parrots: A War Surgeon's Diary*. Milan: Charta, 2005, p. 134.

universal education, he describes "an avalanche of studies, little studying."[2] That is, study after study has shown the importance of universal primary education, even as an AIDS-reduction strategy, but very few children are in school.

We all need optimism, and this book is optimistic. One of its recurring themes is the author's passion for bringing fellow Miamians on-board a crusade to engage his wealthy metropolis, including its Haitian residents and its powerful university, with the long-suffering but proud people of Haiti. Since Fournier and I are both American physicians working in Haiti, and since we work together on certain projects, I'd like to expand on my own views on this crusade. At one point in the text, I am described as "skeptical" about the value of most short-term medical missions to Haiti, and this is true if we're concerned about taking on the most pressing health problems in Haiti and about learning a new way of engaging the citizens of our two countries, the oldest two in the hemisphere.

Let me take each of these two worthwhile issues—which would help us do-gooders do better—in turn. First, how do we take on Haiti's most pressing health problems? Here, short-term medical (as opposed to surgical) missions will often fail, since those who would prevent or treat chronic diseases such as AIDS, tuberculosis, or hypertension need to have a regular presence in the communities served. Also needed are fewer screenings—treatable disease is everywhere—and more follow-up care, and this comes only from determined efforts to rebuild Haiti's public health system and to support our *Haitian* colleagues so that they can serve their fellow citizens. Not with sporadic assistance, but with regular support for community health workers, doctors, nurses, and others. And that is what Medishare, the NGO Fournier helped to found, is now doing in central and northern Haiti; Partners In Health is proud to be a part of these efforts.

---

[2]Lewis S. *Race Against Time.* Toronto: House of Anansi Press, 2005, p. 71.

What of the second project, that of learning to engage with Haiti and Haitians in a respectful and mindful manner? This need is as urgent as the first one; these two issues are in fact intimately intertwined, as Fournier reveals. The book's title is taken not from lurid Hollywood notions of Haitian folk religion but rather refers to the great gift Fournier received from the Haitians, who lifted the zombie curse in question: his own. "I had the zombie curse of a comfortable life in the United States," he writes affectingly.

The notion of zombies is a product of the slave plantation era in Haiti, a time when the French used Africans as things, exploiting cruelly their labor without payment other than the lash, or worse. Haitians say that a zombie is freed if it tastes salt, and Fournier tasted salt in Haiti and in caring for impoverished patients in the United States and in Haiti. Once an American who goes to Haiti is freed from the zombielike condition that keeps us blind about Haiti, once the torpor is lifted, we see the sorry spectacle of our treatment of our neighbors so hard by Miami. From 1804 to the present: over a dozen trade embargoes imposed by U.S. governments; our refusal to recognize diplomatically, until after our own Emancipation Proclamation, a nation born of a slave revolt; gunboat diplomacy; our invasion and armed occupation of Haiti during the first half of the twentieth century; our subsequent support of military dictatorships; and, most recently, our involvement in undermining, rather than strengthening, Haiti's nascent democracy.

To this very day, the impact of these events on medicine and public health is profound.[3] But too many missionaries and short-term medical visitors can't be bothered with these details. How much homework do we have to do in order to do good work in Haiti and

---

[3]Regarding the impact on health of recent embargoes, see: P. Farmer, M.C. Smith Fawzi, and P. Nevil. Unjust Embargo of Aid for Haiti. *Lancet* 2003;361(9355):420-3. The public health ramifications of the most recent coup d'état—Haiti's 33rd—are discussed in: P. Farmer. Political Violence and Public Health in Haiti. *New England Journal of Medicine* 2004;350(15):1483-6.

with Haitians? Does every effort need to be informed by deep knowledge of Haitian history and culture and about our own convoluted
history with Haiti? Since it's hard enough to remember the minutiae
of daily clinical practice, do we really have to know about trade
embargos and the machinations of what are called "the international
financial institutions"? Alas, Fournier's book suggests, indirectly perhaps, that we do. And so do many others who've invested their lives,
their medical careers, in providing care to some of the most desperately endangered people in the world. Dr. Strada, the Italian surgeon
cited above, concludes that he was wrong not to have done all the
homework on what might be called the political economy of the
surgical disease he was seeing:

> We thought that war was an old, primitive instrument, a cancer that
> mankind did not know how to eradicate; on this point we were
> mistaken. Tragically, we—and not only we—had failed to see that
> war, rather than being a burdensome inheritance from the past, was
> becoming a fearful prospect for our future and for generations to
> come. In the operating theatre we saw the devastation produced in
> human bodies by bombs and mines, by projectiles and rockets. Yet
> we did not succeed in grasping the effects of other weapons,
> "unconventional" ones: finance and international loans, trade
> agreements, the "structural adjustments" imposed on the policies of
> many poor countries, the new arms races in richer countries.[4]

Not many readers of *The Zombie Curse* will be interested, a priori,
in "finance and international loans, trade agreements, the 'structural
adjustments' imposed" from above. But if we wish, like Fournier, to
improve the health of the Haitian poor, we'll have to learn much
more about such dismal topics.

In the end, *The Zombie Curse* reads to me as one American's
physician's discovery not simply of Haiti or poverty or AIDS. It's
about his voyage of discovery about human rights and where they fit

---

[4]Strada, p. 132.

in medicine and public health. Seeing health care as a right is a worthy goal, but the path forward is never easy, as this book shows us. It is difficult because ill health is caused mostly by poverty and violence and inequality—and what are we doing to fight those? The more time we spend among the destitute sick, the more we see that the short-term goal of attending to the patient in front of us must be linked to the duty of calling attention to the inadequacy of our own best efforts. The goal of alleviating human suffering must also be linked to the task of bringing others, including those who might read a book written in English, into a movement for basic rights.

Paul Farmer, M.D., Ph.D.
December 2005
Cange, Haiti

Preface

IN THE LATE 1970S AND EARLY 1980S two diseases crept into my professional and personal lives that would change both forever. The first was the newly emerging AIDS virus, a germ that has proven smarter than most of the doctors trying to fight it. The second was one that proved even more difficult to conquer.

Beginning in 1979, as a faculty physician at the University of Miami School of Medicine, working at its affiliated Jackson Memorial Hospital, I saw firsthand the chaos, confusion, and blame associated with the spread of AIDS. Part I of this story examines the early days of the epidemic in Miami. In particular, the story concerns itself with patients admitted to Jackson Memorial, which serves 400,000 of Miami-Dade County's poorest residents.

Nationally, the first groups to be blamed for AIDS were gay men and intravenous drug users. Many old prejudices and stereotypes combined to make these groups easy scapegoats. In Miami, however, a new group of victims emerged. These were Haitian immigrants—proud people who had fled the crushing poverty and tyranny of their island nation for a better life in America. At Jackson the number of Haitian victims dwarfed those of the other groups. A new scapegoat had arrived.

Nothing in my background, education, or training could have prepared me for the misery and mysteries of the early days. I grew up just north of Boston. For all its veneer of cosmopolitan liberality, Boston is arguably the most provincial city in the United States, particularly in the blue-collar neighborhoods of my youth. At that time Boston was de facto segregated. Black people were hardly ever seen in our neighborhood, even though it was poor. Haiti was unheard of, and Haitian people were completely unknown. In college I was vaguely aware of the rise of "Papa Doc" Duvalier and the Ton Ton Macoutes, but Haiti was a minor issue compared to Vietnam, the Cold War, and civil rights. I probably did not even know that Haiti was overwhelmingly black, peopled by descendants of the world's only successful slave revolt.

My immigrant Boston blue-collar upbringing was also quite proper. Openly gay people were even rarer in our neighborhood than Haitians, but it was the AIDS epidemic, more than anything else, that brought homosexuality out of the American closet. Ignorance of the gay experience was a significant contributor to the confusion that physicians, myself included, experienced when AIDS first surfaced in New York City, San Francisco, and Miami. At that time, homosexuality was considered a perversion or a psychiatric disease, not only by society at large but also by most in the medical profession.

Even the vibrant milieu of Miami, perhaps America's most culturally diverse city, was not, in itself, sufficient to understand or deal effectively with the epidemic that was about to overwhelm our community. Looking back, I believe that we of the University of Miami School of Medicine's* faculty fulfilled our obligations to our patients in the traditional manner of medical education—dissecting our patients' signs and symptoms and focusing on diagnosis and treatment.

---

*For most of the time frame of this book, the medical school at Miami was called the University of Miami School of Medicine. The name was changed to the University of Miami Miller School of Medicine in 2004.

We were, however, often naive as to who our patients really were as people, not just as repositories of disease. Unfortunately, we were not the only physicians to struggle against a cultural or socioeconomic gap. This is our system. Few challenge it, and few expect more from their physicians than just to know medicine. In fact, the socioeconomic divisions between doctors and their patients may have been a major factor in allowing the virus that causes AIDS to outsmart us at every turn.

For example, in 1983, I coauthored an article that was later used to label Haitians as "at risk" for AIDS. This stigmatization led to discrimination, an adverse immigration policy in the United States, and feelings of persecution and denial among Haitians. In retrospect, that study was the first clue as to how the AIDS virus would leap from one population (gay men in the United States), to another (men, women, and children in an impoverished country). In fact, in the two decades since that article, AIDS has indeed morphed from a disease predominantly of gay men in the United States into an epidemic that is overwhelmingly affecting poor people worldwide.

The differential burden of an epidemic on the poor is a classic tenet of the epidemiology of infectious diseases. The failure of the public and the professional consciousness to grasp the relevance of this principle to the AIDS epidemic has stymied efforts to contain it. Medical science is obsessed with finding a cure. Because of the tremendous progress toward that goal, for people of means AIDS has become a chronic, manageable disease. But the epidemic continues to rage out of control among the poor.

My generation of physicians grew up in an era of antibiotics, believing that all infectious diseases were conquerable. The ability to cure, which we took for granted, was in truth a limited and transient victory for medicine, with a huge downside: It made us lazy. In reality we can prevent illness, we can diagnose, we can ameliorate symptoms, and we can prolong life, perhaps dramatically. To a certain extent we can predict the future, but we can rarely cure. Even when we can cure, many are left behind and that is an injustice.

The second disease was more spiritual than physical. Part II will show how it took my experiences in Haiti to awaken me to the reality of this metaphysical malady, which I call "the zombie curse," as it afflicted my patients, my profession, and myself. Before going to Haiti, I asked myself the same questions everyone else was asking: Why do Haitians get AIDS? What is it about gay sex that puts gay men at risk? These questions, in reality, were counterproductive. The concept of risk factors had limited utility in the beginning of the epidemic. Now we know the disease is caused by a virus. But the "risk factor" concept persists in the public consciousness, reinforcing the prejudice that the victims are somehow to blame for their own fate or, conversely, that "It can't happen to me." In truth, Haitians aren't at risk for AIDS, and neither are other ethnic groups or gay men, except to the extent that they are exposed to the virus by contact with others who are infected. While, in general, this involves sex with an infected person, there are, as we shall see, exceptions.

In the secret world of HIV—among the homeless, the poor minorities of America's cities and the poor in developing countries— the spread of the disease is inextricably linked to poverty. Interestingly, few scientific studies have addressed the effects of socioeconomic factors on traditional risk groups such as gay men, minorities, or people from developing countries. Moreover, until recently, few in the United States had looked at disparities in health as a function of socioeconomic class as opposed to race. This line of reasoning leads to an important corollary: If we want to win the war against AIDS, medical progress must march hand in hand with socioeconomic progress for the poor and social justice for all "minorities." I laugh as I write this, for those who suffer disproportionately from AIDS and other scourges—the poor, women, gays—are, from a global perspective, the overwhelming majority.

The perception of AIDS as a disease of gays and drug users created a stigma surrounding the disease that even today further complicates efforts to fight it. The vignettes of AIDS victims included here are intended to open your eyes to the suffering and anguish caused by AIDS—not just the physical anguish but also the emotional pain inflicted by ostracism, withdrawal, stereotype, and blame. All of my patients, in their own way, were remarkable. If, as a physician, I could not save their lives, perhaps this recounting of their struggles can make their lives and their struggles more meaningful.

Is there an antidote to this doom and gloom? The answer is *yes*, and it is to be found in the most improbable of places. The story, therefore, is no longer just about AIDS and suffering. It is also about remarkable people who, if we're willing to listen, can show us a way out.

My first trip to Haiti happened almost by accident. Now, more than 100 trips and 11 years later, it seems less accidental. I've wondered half humorously, half seriously, if perhaps a voodoo saint interceded to bring Haiti and me together. At any rate, that first trip was a life-transforming event that evolved into a long-term commitment to the Haitian people. That commitment has, paradoxically, benefited me more than I've been able to benefit Haiti. Finally, I've learned how to effectively fulfill my commitment to serve the poor. For two decades I had been fighting a losing battle against a virus. Now, thanks to Haiti, I'm beginning to learn how to fight back.

Haiti is the poorest country in the Western Hemisphere. This is not the fault of the Haitian people. The reasons for Haiti's poverty can be found in its history and its relationship to Western powers, particularly France and the United States. That history is beyond the scope of this preface, but I would encourage interested readers to learn more about it through books such as Paul Farmer's *The Uses of Haiti* (Common Courage Publishing, 2003). Despite its poverty, or perhaps because of it, Haiti is culturally and spiritually one of the richest nations on earth. Right now, out of that richness, answers to not just the AIDS epidemic but all of the scourges that afflict the poor are being born. Ironically, the colors, contrasts, and contradictions of this impoverished country have come together to teach my students and me more than all of our formal schooling ever could. The mysteries, magic, and faith of the Haitian people have liberated us from our own zombie curse.

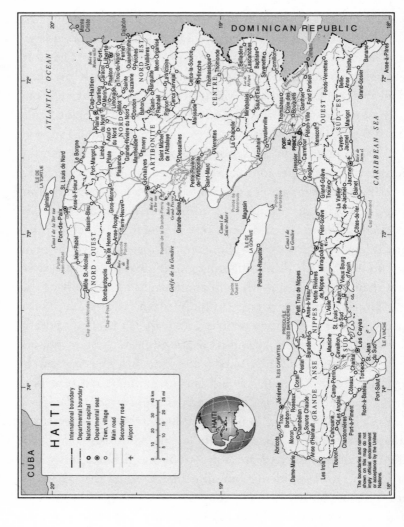

Haiti, Map No. 3855 Rev. 3, June 2004. Courtesy of the Cartographic Section, Department of Peacekeeping Operations, United Nations.

PART I

The Curse Descends

Boat People

1979–1981 "THE NEW ELLIS ISLAND." Miami—Mecca for immigrants and refugees from around the Western Hemisphere. A few blocks north of downtown Miami on Biscayne Boulevard, facing Biscayne Bay and the port, is Miami's equivalent of the Statue of Liberty—the Freedom Tower, Miami's first skyscraper. It is one of the few architectural gems in downtown Miami, registered as a historic landmark, in honor of the hundreds of thousands of Cuban refugees who entered the United States and were processed through this building after the Cuban revolution that started in 1959. From the 1960s through the 1980s, many Cuban refugees became highly successful and influential in Miami. Of course, fleeing communism, the Cubans were welcomed with fanfare. In 1980, 200,000 Cubans arrived in Miami, most during a two-week period that was called the Mariel boatlift. To politically neutralize an attempt to leave Cuba by a handful of dissidents, Fidel Castro sent to Miami all his incorrigible criminals, mentally ill, and incurably sick. Miami absorbed them all. At the same time, an equal number of Haitians were fleeing a brutal dictatorship and misery beyond belief. Unfortunately, the U.S. government branded the Haitians economic, not political, refugees. In addition, there was the small problem of the color of their skin. The Haitians, therefore, arrived in secret—beaching their sailboats, swimming to shore, or escaping into the night from the bowels of freighters docked on the Miami River. If lucky, they connected with family or friends and disappeared into the underground of Little Haiti. If caught, they were transported to the Krome Avenue Detention Center—barracks surrounded by chain-link fencing, barbed wire, and everglades sawgrass, invisible to Miami and the world. For the Haitians, Miami could be a blessing or a curse.

Miami in 1980 was an uncommon city—not to be confused with Miami Beach, the decadent, overbuilt sandbar across the bay. It was founded a little more than a hundred years ago, as much a frontier town as Tombstone or Dodge City. Until the Cubans arrived, it was also very much a southern town. Miami was not integrated until 1967. The downtown area was disappointing then, except to the tourists from Latin America searching for bargain electronics and jewelry. Although magically illuminated, Miami's streets were mostly empty after dark. The life of the city pulsed in the low-rise neighborhoods that surround its center—white and affluent to the south and east, African-American to the northwest, Latin to the west, and Haitian due north.

With a population of roughly equal parts white, black, and Hispanic and significant minorities from the Bahamas, the West Indies, and many countries of Central and South America, Miami is a Creole city—a collection of spices from around the Caribbean and a taste unique unto itself. To most Americans, Miami is a peripheral item as far removed in culture, climate, and geography as it can be and still be attached to the mainland. For its immigrants and refugees, however, Miami is the American dream—the embodiment of opportunity and freedom and the shortest and surest way to flee whatever they are escaping in their own country.

When poor people in Miami get really sick, they have no alternative but to go to Jackson Memorial Hospital, which is supported by Miami Dade County and obligated to treat county residents regardless of their ability to pay. With more than 1,500 beds scattered among several buildings, and with the University of Miami School of Medicine on site, the medical center complex looks like a small city unto itself. I was a resident at Jackson Memorial from 1973 to 1976. After two years with the National Health Service Corps in rural Virginia, I was recruited to return to Miami as a full-time faculty member of the University of Miami School of Medicine. My division chief, Lanny, thought it would be "novel" to have someone on the faculty who had actually practiced medicine. My responsibilities included teaching on the wards and running the medical clinics.

During my brief absence the hospital changed dramatically. Several of the small tile-roofed, low-rise buildings had been torn down and replaced by modern patient care towers. The field of ficus trees filled with starlings that chirped perpetually as I walked from my on-call quarters to the hospital was being replaced by a new ambulatory care center. The hospital relied on the University of Miami School of Medicine to provide its workforce. It was a good marriage. The hospital provided an almost-inexhaustible supply of challenging cases, and the medical school provided a small army of faculty members to supervise the residents who cared for the county's poor. All told, over 500 faculty members and 900 residents worked on the campus.

During my internship and residency at the "Big Jake" in the mid-1970s, Haitian patients were a rarity. Occasionally, a Haitian merchant seaman would be let off a ship and admitted with malaria or a migrant worker would develop dysentery. When I returned to the medical school faculty two years later, I discovered that over 150,000 Haitians had arrived during my absence. Haitian patients were a daily fact of life on our wards and in our clinics. Because of their undocumented status and their poverty, most Haitians tended to wait until they were desperately ill before coming to the hospital. It was difficult to get to know them because of their unusual language, a unique mixture of old French vocabulary with African grammar and syntax. As a group, however, the Haitian immigrants struck me as gentle, friendly, and willing to work at practically anything, particularly if it allowed them to "make it" in this country. There was much discussion about the "Haitian phenomenon" among my colleagues and friends. Most believed that only the brightest and those most driven to succeed were able to gather the resources necessary to allow them to escape to Miami. Life supported by a menial job in this country was superior to a "middle-class" existence in poverty-stricken Haiti, the poorest country in the hemisphere. However, few of my colleagues knew any Haitians, and my knowledge of them remained superficial, limited only to what was necessary to meet their medical needs.

There was discussion at work as to whether they really sailed from Haiti to Miami in their handcrafted, open sailboats. Some thought they flew to the Bahamas and then were crammed into boats at so much per head and towed across the Gulf Stream by organized smugglers. When pressed, however, the people who felt this way could produce no proof other than incredulousness that anyone could survive a journey under such adverse conditions or the fact that their clothes were always so neat and pressed when they arrived. The Haitians later told me that their tradition was to always bring a carefully wrapped and protected change of clothes to mark their arrival and make a good impression in America. Ironically, those clean, neatly pressed clothes often reflected the colors that characterize one of the island nation's most intriguing contradictions—blue and white (symbolic of Christianity), and red and black (symbolic of Voodoo).

At the time, I kept my sailboat moored at Crandon Park on Key Biscayne. For a time the Coast Guard kept all confiscated Haitian vessels there, lashed together two or three to a mooring. I would have to row past them in my dingy to get to my boat. Although the boats themselves were handsomely crafted examples of folk art, to look at them it certainly seemed incredible that they had survived an ocean crossing. They averaged 35 feet in length, with brightly painted wooden hulls, fine lines, broad beams, and hand-hewn tree trunks for masts. Landing on our beaches, each boat carried 60 to 80 people. I simply could not imagine that many people traveling 700 miles in an open boat. One day I rowed over to one for a closer look. Peering over the gunwales, I stared at the inside of the hull stained with the grim but unmistakable colors of diarrhea, blood, and vomit. *What must drive these people to risk such an ordeal?* I wondered. *And how many people have died completely unknown in the passage?*

By early 1980 I sensed that something strange, different, and desperate was happening to some of our Haitian patients. Before 1980, spectacular illnesses caught our attention for teaching purposes like an exploding firework and faded just as quickly. Gradually, however, several faculty members simultaneously realized that what was happening to some Haitians was different from the rare, sporadic illnesses that not infrequently presented themselves to our training program. The patients with these diseases were "classics," behaving in accordance with the classic laws of medicine. But there was nothing classic about what was happening to the Haitians. Some of their illnesses defied all the rules, both in their severity and in the manner of presentation. At first we ascribed the severity of their illnesses to familiar problems in developing countries—poverty, malnutrition, and tuberculosis. By early 1980, however, there were just too many facts emerging to let us continue in our complacency. The illnesses were so bizarre that Lynn, the chair of family medicine, invited a Voodoo priest to consult with him. "It can't hurt," he explained, "and some Haitians truly believe they have a spell cast on them." Word of this consultation spread rapidly through the medical center. While Lynn already had a reputation as being something of an eccentric, the Voodoo priest episode only cemented that impression among most of his peers. To me the patients with these strange illnesses seemed terribly frightened, and medical science had little to offer them to assuage their fears. Lynn may have been onto something after all.

The first person I met who, in retrospect, had acquired immunodeficiency syndrome came into the general medical clinic on a Wednesday afternoon sometime in the fall of 1979. One of the residents who was scheduled to see patients that afternoon fell ill unexpectedly, and I was helping out by seeing some of his patients.

One of them, Jean Baptiste, had just been discharged from the hospital two weeks earlier with the diagnosis of tuberculosis. Since we had effective treatments for tuberculosis, he was supposed to be getting better, but he wasn't. He was emaciated, he could barely walk, and when he did, he fell to one side. His face drooped on the left, and he had signs of spasticity, which indicated that something was the matter with his central nervous system.

When I told the senior emergency room resident that I was sending a patient down to be admitted, my presentation was greeted with the kind of skepticism that only a hectic day in our emergency room could generate.

"You've got to give him time to heal, Art. You've got a biopsy-proven diagnosis. He just needs enough time to get better."

When I insisted that he be admitted, the resident finally gave in and said: "Okay, send him down and we'll take care of him." One week later Jean Baptiste came back to the clinic, having been discharged from the emergency room. By this time he was too weak to walk and had to be assisted by friends. This time I forced his admission on the emergency room with the threat of disciplinary action.

Jean Baptiste was admitted to Lanny's ward team. Lanny, in addition to being my division chief, was also the director of the training program. In those days we used to carpool to work together. While riding into work the following morning I mentioned the case of Jean Baptiste as an example of the tribulations of junior faculty dealing with know-it-all senior residents. Returning home that evening, Lanny informed me that Jean Baptiste was certainly gravely ill and Lanny wondered about the possibility of tuberculous meningitis. Two days later he told me that a CAT scan of Jean Baptiste's brain showed several large lesions. The working diagnosis was tuberculomas of the brain, a rare localized collection of tubercular pus in a place that it normally should not be.

Over the next two weeks Jean Baptiste continued to deteriorate and finally died despite heroic efforts on the part of Lanny's ward team. Lanny's assessment was that he just had too much disease and had come in too late. This seemed like a reasonable explanation at the time. The story of Jean Baptiste became another vignette for me to pass on to my colleagues as we traded stories of the amazing illnesses that were so often cared for on our wards.

Over the next several months these horror stories, traded among the faculty over coffee or between case presentations in the clinic, clearly seemed to involve a disproportionate number of Haitians. Nearly all of them had tuberculosis, not the usual kind involving the lungs, but tuberculosis that spread through the lymph nodes, the liver, or throughout the whole body. We all had cases of tuberculosis discovered in the tonsils or under the vocal cords, by liver biopsy, or by spinal tap. We were amazed that a disease we thought we knew so well could behave so virulently. But it still remained a phenomenon, a spectacular disease caused by a particularly virulent strain of tuberculosis, perhaps compounded by malnutrition and living in close quarters. Many would temporarily get better with treatment. Some developed strange neurological symptoms or superimposed pneumonia. These patients invariably died.

I don't remember who among the faculty first suggested that we group together to study the problem. We all shared a vague sense that something important was going on. Little did we know it would be the greatest medical mystery of our lifetimes. With the exception of Gordon from infectious diseases, we were all from general medicine. Danny and Mark worked closely with me in our particular part of the training program. Art P. was older than the rest of us and had a background in pulmonary medicine, before joining general medicine two years previously. Margaret was junior to Art P., Gordon, and me, but soon took an informal leadership role in the project. Robby, one of our chief residents, rounded out the group.

Soon we were meeting on a regular basis. Already we had each come to the independent conclusion that disease fostered by poverty and neglect did not completely explain the remarkable illnesses we were seeing. These patients were men and women from various walks of life—housewives, students, migrant laborers. The only thing they seemed to share was the fact that they had emigrated from Haiti. We all had theories. I was most interested in why tuberculosis was not confined to the lungs, as is usually the case, but rather spread throughout the body. Art P., who knew a lot about tuberculosis, was not impressed with this. He thought disseminated tuberculosis was always more common in younger populations. Danny was interested in how malnutrition might explain what was happening. Mark and Gordon were more interested in the various kinds of unusual infections.

We decided to review all Haitian admissions during the previous year according to a clinical research protocol. In addition to tuberculosis, several unusual infections were documented. Strangely, a single patient frequently had more than one unusual infection. Some patients received an autopsy after they died. We were surprised to find that what we had assumed to be tuberculosis was frequently a parasitic infection of the brain. Suddenly, our pathologists became interested.

New information began falling into place. Fungal infections of the central nervous system, other infections that usually affect only cancer patients receiving chemotherapy, and disseminated viral infections were discovered by our review. My focus on the question of tuberculosis clearly was too narrow. Our patients were behaving as if their immune systems weren't working. Our suspicions grew further when the pathologists began reporting pneumocystis in biopsy and autopsy specimens. Pneumocystis is an amoeba-like organism that only infects patients with weakened immune systems. The Haitian AIDS mystery began to unfold.

Morts et Mystères*

BY EARLY 1981 OUR REVIEW of the Haitian patients admitted during the previous year showed that the men and women who developed unusual infections were young, had lived in the United States for several months to a few years, were not malnourished before they became ill, and worked in a variety of occupations. They all had blood tests that showed they had previously been exposed to several viruses, including hepatitis, and frequently, but not always, the germ that caused syphilis. Among those with disseminated tuberculosis, a majority had yeast infections in their mouths and swollen lymph nodes in several parts of their bodies. The list of infections we discovered was impressive: viral infections of the esophagus, disseminated herpes virus, disseminated fungal infections, central nervous system fungal and viral infections, central nervous system parasitic infections (toxoplasmosis), and *Pneumocystis carinii*. Frequently, two or more of these infections were found in the same patient.

It was Margaret who first made the connection between what we were seeing in our Haitian patients and the recently reported occurrence of opportunistic infections in previously healthy homosexual

---

*Literally, "The dead and the mysterious." In Voodoo, the phrase refers to the spirit world.

men. Once pointed out, the similarities were indeed striking, but the two groups were not completely alike. First, we had heterosexuals, including women. Second, our patients had much more tuberculosis and toxoplasmosis and much less pneumocystis pneumonia. We had only one case of Kaposi's sarcoma, a previously rare cancer emerging in the gay population. Still, we knew we were on to something and began meeting weekly.

Margaret remarked that the problem was being reported more and more in gay men. "It even has a diagnostic category for billing. They're calling it the Gay-Related Immunodeficiency Disease."

We decided to follow all new Haitian admissions. Margaret and Art P. drew up a questionnaire, and each of us took turns for a week identifying and reviewing all the Haitians admitted to Jackson Memorial. Any patients who seemed to have the syndrome would be followed by whoever picked them up during his or her week on call. I volunteered my office as a logical place to see patients after they were discharged. As it was located behind the medical clinics, I could arrange for patients to be seen there, regardless of their ability to pay the university's usual private patient fees. While most of our time as medical school faculty was devoted to supervising residents caring for "public" (that is, poor) patients, each faculty member was required to devote a small portion of time to seeing private patients. Dan, Mark, Margaret, and I already saw our private patients there, and Fanny and Clara, our secretaries, could facilitate appointments and other logistics. It was clear that these patients could not be well accommodated by the hospital's clinic system, with its long waiting list for appointments and inflexible scheduling.

My week on call finally arrived. The word from those who had already taken call was that I could expect about 10 Haitians to be admitted during the week but that only two or three might actually have the syndrome. The most difficult part would be coordinating my schedule with that of the Creole interpreter. Speaking to patients through an interpreter, in their own language, I began to realize how shallowly I knew these people. During rounds I would introduce myself and try to review important historical points in my best college French. The patients would stare at me blankly or answer in English "Yes, yes" and I knew not a single word I had spoken had been comprehended. Now I was able to examine their lives in exact and intimate detail. Through their language I discovered their intelligence, emotions, sophistication, and sense of humor. My interest in the research took a back seat to my growing fascination with the patients themselves.

I loved them from the beginning. I loved them because they were underdogs. I loved them for their improbable names: Theophile ("love of God," in Greek), Clairvoyante ("fortune-teller"), Marc Aurele (the Roman emperor/stoic philosopher), and Mercidieu ("thanks be to God"). I could trace part of their culture to France, for many of their names—Voltaire and Rousseau, for example—had a hint of the Enlightenment. The language shared with its parent French a rhythm and softness and seemed to have an intrinsic rule that it be spoken while smiling, no matter how much the speaker was suffering. At the same time, I sensed in their speech an Africa of long ago. It sounded like French but was incomprehensible as such and was peppered with repetitive sounds, almost as if it were intended to be danced to, with drumbeats as accents. Their families, particularly, their children, were dressed in a way that surpassed style and approached artistry, even though they were poor. These features of custom and language gave the Haitians an exoticness that enhanced my attraction to them. Poor, peaceful, humble, and hungering after justice, they seemed to be the beatitudes personified.

The housestaff were not quite as sanguine in their opinions. They gave whatever was necessary in terms of hours, dedication, and compassion to these patients with overwhelming illnesses. But already Haitian admissions were getting a reputation as "bad hits," and a sort of gallows humor was beginning to emerge. During rounds one day, in response to an uncommonly prejudicial remark by a medical student, I remarked that I had rarely seen a Haitian admitted with any of the diseases we usually associate with alcohol or drug abuse.

One of my interns retorted, "That's because they don't live long enough."

I interviewed a 42-year-old woman named Marie. It was hard to believe she might have the syndrome, despite her disseminated tuberculosis and telltale oral yeast infection. She looked younger than her age and was slightly overweight. Two years before, she had left her six children in Haiti and come to this country to do domestic work, sending her meager earnings back to her family. She smiled incredulously when I asked through the interpreter if she had ever had a bisexual lover or sex with a woman. Questions about oral sex and anal sex were met with the same look of surprise. Yet she was not offended by the questions and answered in a matter-of-fact manner. She had several boyfriends in Haiti before the birth of her first child, but was then monogamous until her husband died and she came to this country. There had been no unusual sexual practices, just ordinary relations between wife and husband.

The only other patient I picked up that first week was named Claude. He was in his mid-20s and had come here as a student. Again, I saw the same incredulous smile and disclaimers in response to my questions about homosexuality and sexual practices. He looked much sicker than Marie; he was wasted and suffering from high fevers. Still, he was polite and agreeable and seemed glad that someone was taking an interest in his illness.

Blood was drawn, and my forms were completed and then passed on to Margaret, who coordinated things, along with my assessment of "one probable, one definite." Claude continued to decline in the hospital and died of toxoplasmosis a few days later. Marie was discharged in reasonably good condition; however, she did not return for her follow-up appointment. Two months later she arrived in the emergency room with overwhelming pneumonia and died within 24 hours. Margaret informed me that my "probable" had become a "definite."

Haitian patients with the syndrome continued to be admitted to my ward team. Previlus presented with fever, diarrhea, and disseminated tuberculosis. He was a slight man, smaller than average but muscular. His hairline had receded, and he kept his hair trimmed close to his scalp. He had somehow found his way to us from Palm Beach County, where he lived and worked as a migrant laborer. He spoke articulate English and French, in addition to Creole. He was the first to complain to me of itchy bumps on his skin, and we asked dermatology to see him. The consultant's diagnosis was flea bites, but when informed of this opinion, Previlus protested adamantly.

"I don't have fleas, Docteur."

"I understand, Previlus, but that's what the skin specialists think."

"I am not a dog, Docteur."

"Perhaps they're some other kind of insect bite. You do work in the fields. Perhaps red ants."

"I have never had these before, Docteur. There are no insects on me. I have no fleas."

I let the issue die, not knowing he was right. The cause of his diarrhea was discovered through the persistence of a fourth-year student rotating through the team, who would not accept my explanation of this problem by conventional causes. He discovered an unusual parasite in Previlus's intestines that was not supposed to cause disease in humans. Previlus was the first patient with AIDS in whom this infection was discovered. Unfortunately, none of our treatments brought him anything more than temporary relief. Although we could not relieve his itching or cure his diarrhea, Previlus did improve enough with treatment of his tuberculosis to allow him to leave the hospital. In fact, diagnosing tuberculosis in its myriad new forms and effectively treating it (it took three medicines for at least nine months) was one of our first real successes. The old vernacular name for tuberculosis was "consumption," which graphically described how Previlus and the others with tuberculosis looked when they first presented—gaunt and wasted, as if being slowly consumed by a fire burning inside them. Fortunately, after a few days of treatment, Previlus's fever came down and his appetite improved dramatically. After his discharge, I volunteered to follow him in my office, where he joined Theophile, Marc, and Belony, the original group of survivors.

Theophile was picked up by Gordon when he was admitted with a type of fungal meningitis. When he came to the office he always looked remarkably well, wearing a brightly colored shirt and a broad-brimmed straw hat. He was tall and thin but not wasted and had an infectious grin. He complained of headaches after his meningitis, and Gordon treated him with codeine. Whenever he ran out of medicine he would show up unexpectedly at our office. Since I was there more often than Gordon, I would frequently renew his prescriptions.

Marc was one of the first in whom we diagnosed toxoplasmosis of the brain before he died. Initially he responded dramatically to treatment, but the nursing home he was discharged to inadvertently discontinued it. When he returned to us he was paralyzed on his right side and could not speak. Although restarting antibiotics forestalled his death, the drugs did not restore his strength or his speech. He returned to our office in a wheelchair each week to see Margaret. He smiled on one side of his face and drooled on the other in response to greetings from Fanny and Clara. They mercifully ignored his disability and carried on one-way conversations with him: "Oh, Marc, you're here. The doctor will be right with you. You look like you're doing better."

Belony, like so many of these patients, was a student in his early 20s who lived with his mother. His English was not as good as Previlus's. He was more debilitated than the others and more frightened. Rumor had it that he believed he was hexed. His mother, who brought him to each visit, clearly thought so. She adorned his neck with a makeshift amulet, a little sack containing herbs, miniature pictures of saints, and strange hieroglyphic drawings. His hair was falling out in patches, and he covered his head with a stocking cap. As Belony had little spontaneous speech, his mother would answer my questions. He avoided eye contact, preferring the floor. I felt uncomfortable about not having the same rapport with him that I had with the others, but medically, at least, he was not getting worse. In fact, he remained as I have described him for several months.

Despite our patients' marginal health, I was optimistic. It was exciting. The veil of ignorance had been partially lifted. Patients who would have previously died were surviving and leaving the hospital. Many did reasonably well between relapses, despite their blood tests, which showed that they were still immune deficient. We had effective treatments for many but not all of the infections and hoped that if we could buy enough time either we would find a cure or the patients would recover spontaneously.

We also knew something that hardly anyone else knew: This terrible disease did not affect homosexual men exclusively. The speculations in the letters section of the *New England Journal of Medicine*—amyl nitrates, proctofoam cream, the immunosuppressive properties of sperm in the bloodstream—were fanciful but wrong. But what was the connection between these Haitians and gay men with the same illness? Was this something new, or had it been there all along but in our ignorance we had missed? We speculated among ourselves. New virus? Mutant virus? Combination of viruses in sequence? Genetic predisposition? Old virus behaving in a new manner? Perhaps it was exposure to malaria or some other parasitic organism endemic to Haiti? There certainly were enough possibilities to choose from—leishmaniasis, schistosomiasis, strongyloidiasis. Rifampin, a drug used to treat tuberculosis, was reported to have immune-suppressant properties. Each of these theories was considered, but there were no hard facts to support any of them.

Mary Jo in obstetrics delivered the baby of a pregnant mother dying of tuberculosis. Gwen and Wade in pediatrics followed this child and soon others, some born of seemingly healthy mothers. All eventually succumbed to bizarre infections. So now we had infected men, women, and children in growing numbers. Their only link? They were all Haitians.

Régis

DAN AND I ALTERNATED MONTHS as attending physicians on the in-patient service. I was about to take over responsibility for the service and Dan was briefing me on the patients. We had been working together for three years and frequently commiserated about the plight of our patients.

"The sickest is definitely this fellow Régis. Have you heard of him? He was admitted with *Pneumocystis carinii* pneumonia. His blood count started to fall on Bactrim. We stopped it, and we're waiting for Pentamidine to arrive from the CDC (Centers for Disease Control and Prevention). He's dying fast. You may be forced to restart the Bactrim. It's remarkable. He was a dentist in Haiti."

When I met Régis on rounds the following day he was near death. He had pneumonia throughout his lungs. He was breathing heavily at three times the normal rate and was too weak to talk. His mouth was dry, despite the oxygen mist streaming from a mask over his face, and his eyes were rolled back in his head. When he was admitted his temperature was 105°F. It decreased to 102°F while on Bactrim but climbed again when his antibiotic was changed. Because of the glut of terribly sick patients the syndrome was already causing, there were no beds available in the intensive care unit. My resident and I decided we would give the pentamidine one day to work, watch Régis closely, and then intubate and artificially ventilate him if necessary. We would not let the absence of an intensive care bed keep us from doing everything we could. After questioning the medical students and answering their questions, my team moved on to the next patient. As sick as he was, Régis was only one of about 25 patients under my care.

The next morning Régis was even worse. The intern on call had been up all night with him, restarting intravenous lines, drawing cultures, and frequently checking his arterial blood gases. His chest x-ray showed more consolidation of the pneumonia, and his blood gases were deteriorating. We decided to give up on the pentamidine and restart the drug he had been on three days before. Bactrim could kill the organism causing Régis's pneumonia more effectively, but it had stopped his bone marrow from making red cells. Now this seemed the lesser of two evils. We could always transfuse him, and we hoped a special vitamin-folinic acid would reverse the drug's effect on his bone marrow.

The intern on call was again up all night ministering to Régis. By the following morning he showed signs of improvement. He was breathing more easily and his chest sounded clearer. His temperature had dropped to 101. He had the strength to talk again.

"Who are you?" he asked me as I leaned over to listen to his chest.

"I'm Doctor Fournier. I'm in charge of the team you were admitted to."

"Oh, the name on the bracelet?" he smiled as he pointed to the identification band on his wrist that contained his name, his hospital number, and my name. In a teaching hospital hardly anyone notices the name of the attending physician on the I.D. bracelet.

Even after having so recently climbed out of the grave, he was handsome and noble looking. His skin was truly black, unblemished, and shining with the moisture of perspiration and the oxygen mist. His eyes were animated and accentuated by angular cheeks and a broad, sharply crested nose. His teeth were impossibly white and perfectly shaped and spaced. His English was perfect, without a hint of an accent.

There was no doubt he had AIDS. In addition to pneumonia, he had patches of fungus inside his cheeks and on his palate. The number of lymphocytes in his blood counts was depressed, a sure sign, and his "helper" to "suppressor" T cell ratio was inverted. T cells are the part of the immune system that fights off unusual infections. We now know that the AIDS virus specifically attacks the "helper" or "T4" lymphocytes slowly, over time depleting their numbers. When they reach critically low levels, patients become sick with unusual infections. But Régis had no signs of tuberculosis and a test for syphilis was negative. The residents were justifiably proud of pulling him through, especially since they had done it without the benefit of an intensive care bed. Paul, his intern, couldn't wait for my arrival at rounds the next day.

"Boy, you're not going to believe how much better Régis is today. The man's incredibly smart. He was asking all about his illness, and I told him about the pneumonia, and the Bactrim and the pentamidine and how his immune system is all screwed up. He understands it all. His blood gas is almost normal, and his chest x-ray's even starting to look better."

When I arrived in Régis's room, I was greeted with, "Good morning, Dr. Fournier. How are my T cells doing this morning?" He was up in a chair, had washed, and was able to breathe comfortably without oxygen. I paused for a moment to let the residents bask in their accomplishment.

"Well enough. You have improved greatly."

"Then you think I will recover?"

"You're recovering already."

"But will my T cells recover?"

I would have taken this last question as a joke except that he asked it with complete sincerity. I had never been asked a question of such immunological detail by one of my patients. Clearly, he was seeking reassurance for questions deeper than T cell function. I felt constrained by the format of attending rounds and answered with a trite, "Only time will tell."

I shared the resident's exhilaration in pulling Régis through. He had been all but dead three days before. Had he presented as little as six months earlier he surely would have died despite our efforts, as we wouldn't have had a clue as to the true cause of his pneumonia or the best treatment. His strength increased daily. Soon we incorporated Régis into our rounds, asking him to translate for all our Creole-speaking patients, rather than waiting for an interpreter. He was excellent at this, reporting not only what the patient said but also an assessment of the patient's level of understanding and unspoken concerns. He added a touch of drama and eloquence as he told each patient's story. The patients were puzzled at first as he emerged from the group of physicians and students wearing a hospital gown and pajamas and still attached to an IV pole. But he stated plainly what he was doing and then put them at ease with a smile and a handshake. We all remarked what a luxury it was to have our own interpreter and one of such quality.

Régis seemed different from many of the other Haitian patients. For one thing, no family members ever visited him. He was always reading or writing. His bedside table had only a King James Version of the Bible to adorn it, rather than the usual pictures of saints that graced the tables and walls of other Haitian patients. "Mysterious," I thought.

One day toward the end of his hospital stay I was making rounds alone, dictating my daily notes into a handheld recorder. When I got to Régis's room I decided to stop in, since I had missed the opportunity to see him during particularly hectic morning rounds with the team. I knocked and entered. He was reading his Bible.

"Hello, Régis. I didn't get to see you during rounds this morning. Continuing to make progress?"

"Yes, doctor. I think my lymphocytes are holding their own now. Thank you. Won't you visit for a while?" He closed his Bible and placed it on a photo album on top of his bed stand. I accepted his invitation and asked how he had come to this country. He told me he had grown up in the Haitian countryside but had always been a good student at the Protestant missionary school in his village. After high school he took a correspondence course in dentistry, there being no dental school in his part of Haiti, and the mission sponsored him in setting up and running a dental clinic. He did this for five years and even wrote a book for the public on dental hygiene. As he told me this, he pulled the book from his drawer and displayed it with a smile. Now he had come to this country to formalize his education. He both studied and taught English as a second language at a local college. He was 33 years old.

The book was in English and was obviously written at a time when Régis's English had not reached its present state of perfection, for it was grammatically but not idiomatically correct. His picture was on the frontispiece, handsome in a three-piece suit. The content concerned fundamentals of caring for teeth—the importance of brushing and cleaning and what happens during a visit to the dentist. It had been published by the mission. He was quite proud of it.

The photo album, however, was of greater interest to me. There were many pictures of Régis surrounded by white people somewhere in the United States. These were his visits to the mission's stateside base. Interspersed with these were photos of him in an obviously tropical setting, usually surrounded by groups of children or adults. But one picture arrested me—Régis in a small, bare room with one chair and an older woman seated in a handmade chair. Régis had one bare hand in her mouth and with the other was extracting a tooth with a pair of pliers. The woman's face was contorted, and her legs were crossed in pain. Remembering cases of dentists with hepatitis, I could not escape the revelation: "This was how he had contracted the disease!"

I told him how fascinated I was by the pictures of his life in Haiti. He seemed pleased and told me I should visit Haiti someday. He asked me what would become of him. It was clear that the knowledge he had absorbed from Paul's lessons in immunology was more than superficial. But he was calm and not frightened. He said he had to get better so he could finish his education and return to his work in Haiti. I told him that his problem was serious but not hopeless. Several patients had died, but more and more were surviving, and this illness was still too new to predict the future. I offered to follow him in my office and promised that, if the situation did become hopeless, I would tell him.

Blood Brothers

Now THE MEDIA HAD DISCOVERED the "gay plague." Every morning the newspaper and every evening the television news had another sensational story. Who could blame them? It was new, it was contagious, and it had to do with sex.

Although our group had gathered convincing evidence that AIDS was not confined to gay men, for a while that remained our secret. Actually, not quite a complete secret. Eventually we notified the Centers for Disease Control of our findings. The CDC sent a task force to meet us, saw some of our patients, reviewed our data, and helped us with special immunological testing. We heard from the agency that some patients from the Haitian community in New York City had come down with same illness. We were preparing our data for publication.

For most of us in the original study group, AIDS among Haitians remained a part-time endeavor. Margaret, however, was devoting more and more of her energy to the problem. She followed the largest number of patients and came to the office to see them practically every day of the workweek, instead of the one-half day a week usually devoted to the faculty private practice. She and Art P. were planning a trip to Haiti to look for evidence of the syndrome there.

She talked about the subject with an urgency and an excitement the rest of us didn't seem to share.

I continued to follow Previlus, Belony, Régis, and a few others. Previlus was doing reasonably well. His skin continued to itch, despite a host of nostrums prescribed by the dermatologists, and his diarrhea flared sporadically. He missed many of his scheduled appointments and then would show up unexpectedly, explaining how difficult it was to get down from Belle Glade. However, he continued to work, his weight was stable, and he developed no new infections. He was grateful that I was caring for him.

Belony was slowly declining, and I could not put my finger on why. Each visit he required more assistance from his mother to get up on the examining table. He said very little. His mother reported that he spent his days lying in bed.

Régis was doing the best of the patients I followed. He felt well, his appetite was good, and he was gaining back the weight he had lost during his bout with pneumonia. He would come to the office in a suit and tie looking like a foreign ambassador. He made light, pleasant conversation with the office staff. For a while he was doing so well that I thought he might be the first spontaneous recovery. But his lymphocyte count remained low, and tests showed that he was still immunodeficient despite his outwardly robust appearance. One bothersome problem was the development on both ankles of the same type of itchy bumps that plagued Previlus. I did not even attempt to offer the dermatologist's explanation of "insect bites" to Régis. Fortunately, since he was so concerned about his appearance, the bumps were hidden beneath his trousers.

Margaret asked me if I knew anyone who had B+ blood type. I answered that I was B+, knowing full well that in answering I was also volunteering for something. I wondered if she had somehow checked in advance. She wanted to mix the lymphocytes and sera from our Haitian patients with "normals" and inject them into marmosets. Régis was B+, and that being a relatively rare blood type, she had yet to find a "control." "It's only 50 cc's," she demurred. So I drew the blood sample from Régis, and Margaret drew a sample from me. "I guess this makes us blood brothers," said Régis, smiling.

All of these patients required lots of attention and we saw them frequently, but the nature of our office waiting room was changing. With the exception of Régis, there was no mistaking that these patients were poorer and sicker than the private patients who shared the waiting area with them. The private patients would consciously or unconsciously sit as far from the Haitians as they could. I'm sure many wondered what they were doing there. If two or more Haitians were waiting, they and their families would chatter in Creole, while the private patients waited silently, with their noses buried in *The New Yorker* or *Sports Illustrated.* Sometimes, if a particularly sick Haitian was waiting, a private patient would stare as if seeing an apparition. Still, no overt objections were raised, for the reason the Haitians were there remained a secret.

Now, two or three new cases a week were being admitted to the hospital. Only half survived to be discharged. One corridor of one floor was exclusively occupied by Haitian patients with AIDS. Because they were poor, without resources, frequently living here illegally and unaware that this disease was among them, many arrived at the hospital moribund. The residents continued to work heroically to keep them alive, but the high mortality was eroding morale.

Leguerre was a case in point. He was admitted with fever and swollen lymph nodes. When presenting him during attending rounds, Jim, his intern, remarked that although his neurological exam showed no localizing signs, his speech did not make sense, even to the Creole interpreter. I commented that even this subtle a change in mental status might be a clue to the kinds of central nervous system infections that afflicted so many of our patients. When we went to his bedside to examine him, we found a patient who had changed drastically since Jim had last seen him a few hours before. He was now paralyzed on one side and unable to speak. Jim was embarrassed at the discrepancy between what he described and what we now saw. There was no need for an explanation, though. I had no doubt that his initial exam was accurate, and Leguerre was not the first patient to deteriorate so rapidly.

So, after being up most of the previous night, Jim began another 24-hour ordeal with his patient. The neurology resident had to be contacted to arrange for an emergency CAT scan. Shortly thereafter seizures developed. The patient then needed to be intubated to protect his airway, and a venous catheter needed to be inserted to give him medicines. Another catheter was inserted to drain his bladder. The intern accompanied Leguerre when he went for the scan, which showed multiple abscesses in the brain.

The following day Leguerre was a pathetic sight. He was lying naked in bed except for the condom catheter covering his penis and the large bandage covering where the venous line entered his body. He also had an endotracheal tube attached to a ventilator coming from his nose. Despite his paralysis, the nurses had restrained his arms and legs so that he would not jeopardize the critical tubes and lines. His eyes looked only to the left, and he was unresponsive. *There is nothing like modern medicine to dehumanize a human being*, I thought.

Several times during the next 48 hours it would have been easier just to give up. Then after two more days Leguerre began responding to the antibiotics. His seizures stopped, so we removed the endotracheal tube and discontinued the ventilation. By the fourth day of treatment his paralysis was improving and he could speak coherently in Creole and even in broken English. Soon he was walking the halls, with only a slightly spastic gait. I congratulated Jim publicly and told him that, no matter what else happened in his life, he could always look back with pride at his vigil with Leguerre. I think Leguerre realized Jim's special efforts too, for he smiled at Jim much more broadly than he did the rest of us. On rounds I used Leguerre as an example of how even the most seemingly hopeless case could be sometimes salvaged with prompt diagnosis and treatment and meticulous attention to detail. But three months later Jim told me that Leguerre had been readmitted and died of an overwhelming infection.

The word was filtering through the hospital that our Haitian patients were suffering from the same disease that the media was calling the "gay plague." The beginnings of the backlash were appearing—transportation workers refusing to escort patients for x-rays; nurses worrying about our patients in semiprivate rooms sharing bathrooms with other patients; interns skipping rectal examinations; surgical residents dragging their feet on performing biopsies and necessary operations.

Back at the office, Margaret had started seeing gay men with the syndrome. Although she made no announcement, there was no mistaking the fact by those of us who shared the office. Some were stereotypically and openly gay. They dressed effeminately and talked with our secretaries as if they were sisters. Others were recognized only because they came back time and again with the same men. Although some were resigned, and all held up under the strain with remarkable dignity and fortitude, most were anxious. Their anxiety was not helped by Margaret's schedule, which frequently had her in two places at once. There were frequent outbursts of anger as they waited. Some with Kaposi's sarcoma had large, red lesions on their arms and faces that announced to the whole world that they were gay and had AIDS, like a modern-day scarlet letter. Some came with their parents. Others came with their lovers. Occasionally couples would know each other and make pleasant conversation about mutual acquaintances or interests. Usually, however, the mood was somber, especially if there were a particularly sick or wasted patient in the group.

I had many discussions about this new development riding to and from work with Amal. She is an Egyptian Christian physician who lived with an American family in my neighborhood and worked in our office doing research with Mark on hypertension. She was fascinated with everything American and was an incurable optimist and a compulsive storyteller. We were in the habit of having wide-ranging discussions about politics, religion, culture, and morals to pass the time as we commuted. She claimed an intense interest in the mysteries of life, which she attributed to her ancient ancestors, the children of Pharaoh. She looked remarkably like pictures of Nefertiti, with large eyes and straight black hair. She often told me it was the Christian Egyptians who were the true descendants of Pharaoh, not the Muslims, who sprang from Arab invaders. She held profoundly fundamentalist religious views, steeled by belonging to a religious minority in her own country. She did not believe in evolution, she interpreted the Bible literally, and she held a Calvinist view of fate. Nothing happened unless God willed it.

She had several vignettes she would recount during our rides to work to illustrate God's active intervention in her life. I was such an intervention—only a beneficent deity would have arranged for her to live in the same neighborhood as me and to have me pass directly by her home on my way to and from work. In exchange for this gift she felt compelled to convert me from skepticism. Fortunately for me, I had had the opportunity to previously practice all the religious arguments as I passed through Catholic high school and college. I therefore took these attempts at conversion as scholastic amusement, while she took them with sincerity and earnestness. On the assumption that the best defense is a good offense, I told her my theory of the homophilic origins of early Christian theology. This theory had been finely honed many years before during free periods in my college canteen. Of course, with all that talk of brotherly love and traveling around with a bunch of guys and never getting married, saying Jesus was gay was an easy first step. But three early followers deserved critical review. Was Judas' behavior that of a lover scorned? Da Vinci thought as much in "The Last Supper." And was St. John not the disciple that Jesus loved? Finally, there was Saint Paul. He was a Grecophile and a misogynist. By interpreting the experience of Christ in the light of Greek thought, he linked the God of Love forever to Plato, who clearly felt homosexual love was love in its purest form.

Such talk both scandalized and titillated her. She was intrigued that an American would even think about such matters. My conversion from skepticism became one of her priorities.

She was fascinated by the closeness of some of the gay couples. "They always come in pairs," she remarked to Fanny.

"Just like Noah's ark," I commented, overhearing her whispers. She claimed there was no homosexuality in her country. "Not even among the Muslims?" I countered. She wavered at this and almost took the bait, since she despised the Muslims, but would not concede the possibility of homosexuality among her countrymen. To Amal, AIDS was a punishment visited on the gays for their sins. Better they should suffer here on Earth than suffer the pains of eternal damnation. The Haitians were more problematic, but she was unshakable in her belief that their suffering somehow manifested God's glory.

I did not tell her about Régis, but she discovered him in the waiting room, probably noticing him reading his Bible. She would frequently talk quietly to him and touch his hands until I called him to the examining room. She often asked me about his progress as we started for home. I could not resist asking her what sin God could be punishing him for. "He is strong in his faith. God will not abandon him," was her reply. Then she paraphrased the story of Job, as if preaching a Sunday sermon.

It was no longer possible to hide the fact that we were caring for AIDS patients. Some of the established private patients started making offhanded remarks to our secretaries, and the clinic administration also began making discreet inquiries into what we were doing. The number of gays was increasing faster than the number of Haitians. Margaret faced rising criticism from private patients, administrators, and unfortunately her peers.

Régis continued to be unsuspected by the casual observer, thanks to his handsome features, conservative dress, healthy appearance, and perfect English. He began complaining of chest pain and shortness of breath, but his physical exam, chest x-ray, and stress test were normal. I suspected he might be sensitized to noticing things that he otherwise would pass off as trivial. Not that I blamed him, but I tried to reassure him that, objectively, he was doing well. He was interested in the results of Margaret's experiment with our blood. Unfortunately, I told him, his blood had killed my lymphocytes.

Zombie

1983   DAN AND I, WHO WERE responsible for our residents' education in clinic matters, asked Jeanette, a young Haitian-American psychiatry resident, if she would talk to the residents about Haitian culture. We hoped these talks might break down some of the barriers that had surfaced between our residents and their patients. Jeanette served as our liaison with the Department of Psychiatry. She usually taught the residents generic issues that general medical doctors need to know about psychiatry—recognizing depression, treating anxiety. The idea of using someone from one culture to teach doctors from another culture was novel. Poised and confident, she set up a television and video cassette recorder and engaged the residents with her eyes.

"Today I'm going to share with you the secret of the zombie curse." Usually it was difficult to get the medical residents involved in behavioral science seminars, but Jeanette captured her audience with her first sentence.

"There are zombies in Haiti, and it's related to Voodoo. Who knows what Voodoo is?"

"That's the religion in Haiti where they stick pins in dolls," volunteered one resident. "People are so afraid of it, that it controls their minds," another resident answered.

"You've been watching too many movies," said Jeanette. "Voodoo is the Creole pronunciation for the French *Vieux Dieux**—the old gods, the spirits of the forest, in Creole the *Lwa*, who can be called out from their homes in the mapou and mahogany trees. The gods the slaves brought from Africa. The French tried to impose Catholicism on their slaves and to a certain extent succeeded. But the old gods, the *Lwa*, continued almost like the Catholic saints—there for personal intervention. In fact, in Haiti many of the saints have two personae—their Catholic image and their Voodoo role. The power of the *dokte fè*—you might call him a witch doctor, but the name really means "leaf doctor"—comes not from superstition but from a refined knowledge of the pharmacological effects of local plants and animals.

Just as the Inuit have many words for snow, the Haitians have several names for practitioners of their secret rites. In addition to *dokte fè*, there's *hougan*, spell giver; *bokar*, a male priest; and *mambo*, priestess. And there are several kinds of spells too; good spells and bad ones, ranging from a *mojo*—a love potion—through curses meant to wreak revenge. The worst, though, is the zombie curse.

Jeannette showed a documentary tape from the BBC of people who were declared dead, buried, and then turned up alive. In one case, a man returned to his sister's house 14 years after he was buried. The residents were spellbound. "The key to the zombie curse is tetratotoxin, found in the skin of puffer fish, abundant in the waters surrounding Haiti. It induces a state indistinguishable from death. The *dokte fè* returns after the funeral, exhumes the body, and administers an antidote that keeps the zombie in a drugged state. Zombie is Creole for 'like a shadow.' Literally, when you're a zombie, you're only a shadow of your former self.

---

*This was Jeanette's explanation of the word "Voodoo"; many scholars believe it is derived from the Fon word for "spirit"—"Vodoun."

"It's a life of slavery, the worst fate possible for the descendants of slaves. The curse is therefore only applied by a secret village council to individuals believed to have violated the rules of society and who need to be cast out. It was a way of keeping the old religion and indeed the old African society alive, and it served as a potent weapon against their French masters. Think of the symbolism. The *dokte fè* has actual power over life and death. You die, and he raises you up, not just your soul but your body as well. That's tough for other religions to match."

"Great lecture!" I thought to myself. "One of the best I've ever heard." Unfortunately, the audience consisted of only myself, Dan, Amal, and six residents. The residents were enthusiastic and congratulated Jeanette for an excellent talk. "Now that you know about the zombie curse, don't keep it a secret. Tell your fellow residents," she advised.

On the ride home I asked Amal what she thought of Voodoo and the zombie curse. "Only God can give us everlasting life," she responded. "It's magic, black magic, a trick. A clever trick, but a trick all the same."

"I think you missed the point," I answered, looking at her and negotiating traffic at the same time. "It's not a question of whether it's magic or real, although that guy who walked into his sister's home 14 years after being declared dead sure seemed real to me. It's a question of whether Voodoo is a real religion or not, on the same par with Catholicism, Buddhism, Islam, and all the others. I think Jeannette made a compelling case that it is a coherent set of beliefs and practices devised to make sense out of the unknown and perhaps to exercise some degree of control over it. The thing about the zombie curse that's so interesting to me is not the spell itself, but the fact that it has the full weight of a law. It's cast as a formal judgment of innocence or guilt." Amal fell silent for the remainder of the ride home. I think our theological discussions were starting to frighten her.

Thanks to the Centers for Disease Control, by the time our paper had been accepted for publication the news had already been released to the media that Haitians were at risk for AIDS. Our local newspaper had a front-page article entitled "Haitians Dying of 'Gay Plague.'"

Other stories suggested that AIDS, in fact, had originated in Haiti and perhaps was related to secret Voodoo ceremonies involving the drinking of blood. These stories, usually accompanied by a picture of a Voodoo priest or priestess slaughtering a chicken or goat, I found particularly offensive. The implication was that one could get AIDS from drinking animal blood or that there were secret Voodoo ceremonies involving cannibalism, vampirism, or human sacrifice. Even having just begun the process of getting to know a few Haitians as real people rather than media caricatures, these stories infuriated me. I could only imagine how the Haitians felt.

Margaret appeared on television several times attempting to explain what we had discovered, but the media always seemed to edit her meaning. As an "insider," it hurt to see how our work was being distorted and misrepresented in the press. Our discovery of 22 patients with AIDS was portrayed as a rising epidemic threatening to engulf the 150,000 Haitian immigrants residing in South Florida. The possibility of heterosexual transmission implied by our data further fueled the flames of sensationalism.

Spokesmen, including physicians, in the Haitian community were particularly upset with us. To a certain extent they were justified. Life for Haitian immigrants in Miami was difficult enough without having the entire community accused of introducing a modern-day plague. Haitians were both fired and not hired because the research performed by us played to the inherent bigotry of some employers. The Haitian community responded to this threat to its existence in our country by accusing us of being bad scientists. We hadn't been able to talk to our patients in their native language; we didn't understand the Haitian cultural taboos against homosexuality and therefore the reluctance of our patients to admit to such practices; we had failed to involve the Haitian community in our study.

Most of this anger was directed at our most visible representative, Margaret.

Our Haitian critics charged that labeling Haitians as being a risk group for AIDS was just a new sort of racism. Each of these accusations, but particularly the last one, hurt us deeply. We had started our project with no preconceived ideas, only the problem of sick patients who happened to be Haitians who were dying on our wards. We never claimed that patients got AIDS because they were Haitian, only that the disease was present in a group whose only apparent link was a common ethnic background, similar to sickle cell anemia among American blacks or ulcerative colitis among Jews. These were medical facts, not political statements; they were clues to the riddle, not bullets for the gun.

In addition to hurting us, these charges also raised the first seeds of doubt about what we were doing. In retrospect we had been naive in our interviewing techniques and our assumption that patients, speaking through an interpreter, would reveal intimate details to physicians they had never before met. We probably did not identify some who were gay, or who had sold themselves in prostitution, or who had visited prostitutes but were ashamed to admit it. But for the most part, those men that we followed over time who told us they were not homosexual seemed quite credible. And what of the women? How did they get the disease if homosexuality was the only risk factor? In attempting to protect the Haitian community from bigotry, our critics were forced to question not only us but also the truthfulness of each of our patients. It seemed to be a no-win situation.

Ironically, in 1983 and 1984, Dr. Luc Montagner in France and Dr. Robert Gallo in the United States respectively, discovered that AIDS was caused by a virus. That discovery made the whole concept of risk factors irrelevant. AIDS was an infectious disease, nothing more. But it was too late. The labeling, stereotyping, and blaming had already taken too strong a hold.

The politicization of AIDS in the press also had adverse effects on day-to-day life on the hospital wards. Panic was spreading among the hospital personnel. It seemed as if everyone was worried about being infected. Some nurses refused to bathe patients with AIDS, transportation workers refused to move patients with AIDS, and in some cases residents avoided caring for AIDS patients. I would be embarrassed to enter elevators and find patients who obviously had AIDS going to the x-ray department or some other part of the hospital being transported by people wearing surgical gowns, masks, caps, gloves, and booties. In defiance, I made an effort to talk to these patients if I knew them, wish them well, and touch them as we parted company. Walking down the corridors I would overhear nurses' assistants saying to each other things like, "I don't care if they fire me. There is no way I'm emptying so and so's bed pan." I made a point to go up to these groups and tell them not to believe what they were reading in the newspapers. I had touched these patients in the course of examining them as much as anyone, going back to before we even knew the disease existed, and I was still alive and well. Usually my pleas for rationality were met with quiet disbelief.

In the beginning, the hospital gave each patient with AIDS a private isolation room. This was one of the few "perks" associated with the disease. Two years into the epidemic, this policy had to be abandoned, as every private room on our floor was occupied by a patient with AIDS. There were concerns about putting AIDS patients in the same room with non-AIDS patients. There were concerns about putting AIDS patients in the same room together because they might infect each other with opportunistic infections. There were concerns about putting gays with AIDS in the same room for fear of illicit sexual activity. These discussions took place not once but with ever-increasing frequency as the number of patients continued to climb. Eventually economics won out. Any qualms about who a patient's roommate might be had to be abandoned in the face of the overwhelming number of patients who needed beds.

Finally, there were fears about taking patients with AIDS home. Families of victims couldn't cope, either because of the overwhelming debility of the patients or because of fear of contagion. Boarding homes and nursing homes frequently refused to take those who were too ill to be cared for by their families. Our wards became "God's waiting room." One poor patient with Kaposi's sarcoma spent four months waiting to die on my ward service, simply because he had no place else to go. During all but the final week of that hospital stay he was alert and capable of caring for himself.

Pressure was increasing from two directions to try to get me to move Margaret's practice. From the clinic administration I would hear about alleged complaints from the private patients about sharing the waiting room, examining rooms, and bathroom with AIDS patients. Are the examining rooms wiped down with alcohol each time after they're used? Are the toilet seats disinfected? And thank God we have disposable speculums! These objections were easy to deal with. I told the administrators that there was no evidence a person could get AIDS from just sitting where someone with the disease had sat. Margaret was taking care of a problem no one else was willing to face and that was not going to go away. Yes, I agreed, thank God for plastic speculums. (It's hard to believe now, but disposable speculums were still an innovation at Jackson Memorial Hospital in the early 1980s!)

It was more difficult for me to deal with complaints from fellow faculty. They had known of our study from the beginning. They were friends. They were working for the same goals that I was. They cared about our patients. They told me that this was no way to run a private practice. The AIDS patients were never going to be a central part of our job, which focused on teaching general medicine. Private practice development was more important. If I wanted the private practice to succeed, the AIDS patients would have to go. One faculty member, who is usually calm and unflappable, was particularly upset. "This office has become Miami's answer to the Turkish baths. Fanny found two gays fondling each other behind the door to her office. How would you feel if you were a private patient sitting in our waiting room and were surrounded by all these homosexuals?

And Theophile, he's walking around begging for cab fare home. You have to talk with Margaret." I told him I would speak to Margaret, but then I stalled.

Meanwhile, Margaret's composure was beginning to unravel. Although she wouldn't admit it, she clearly had more work than one person could manage. Patients sometimes arrived without appointments, having heard about her through the grapevine. They waited all afternoon just to have the opportunity to talk with her and try to make an appointment. Others with appointments waited all afternoon only to be turned away without seeing her, as she was tied up in the hospital ministering "last rites" to one of her dying charges. The housestaff's gallows humor said if you're a patient, you know you are in trouble when they call in Margaret as a consultant. Some very cruel colleagues referred to her as the "fag queen." These sorts of comments, coupled with the criticisms directed at her from the Haitian community, needed only the extra burden of the death of one of her patients to bring her to the verge of tears. On occasion she would blow up at the secretarial staff and then, regaining her composure, apologize. Gwen in pediatrics and Mary Jo in obstetrics were under similar strains—solitary figures fighting a lonely battle against a disease few of their peers wished to face.

Belony declined to the point that he had to be admitted to the hospital. My patient was no longer even recognizable as the person I had started to follow several months before. His hair was almost completely gone, his eyes were sunken and defeated, his skin was excoriated, and his lips were cracked. He was too weak even to acknowledge my presence. He died a few days later. I had followed him for seven months, yet I could not honestly say I had done anything for him. There had been a barrier between him and me that I had not been able to break down. I was not sad at his death. His decline had been so pathetic that death was merely the end of his suffering. But I was disturbed with the possibility that, medically or humanly, I had missed something.

Previlus arrived in my office or in the emergency room approximately once a month with fever, diarrhea, and dehydration. At the time of each admission he looked desperately ill, but he responded quickly to intravenous therapy. After recovering, he would disappear again to Belle Glade. I was impressed with his resiliency. He was always eager to return to work, for he was sending most of his earnings to his family in Haiti. Occasionally he would show up in the office for refills of the medication that controlled the fungal infection in his mouth. On these visits he would always seem to be in a hurry. He would only complain about his itchy skin and inquire whether I had anything new to try for it. Although he knew he had AIDS, I marveled at his nonchalance. He was always so concerned with mundane and trivial matters, despite the impending doom. The way he would materialize without warning either in the office or in the emergency room contributed to my growing feeling that things were getting out of control.

I had to admit Régis once with a recurrence of pneumonia. This time, though, he was not nearly as ill. His hospital stay was uneventful. Amal visited him in the hospital. Afterwards she asked me if there was any hope of recovery. I told her truthfully that we had yet to see a spontaneous recovery but that I still had hope for Régis since he and Theophile were doing the best of all the Haitians we had followed. I did not tell her that I was disturbed when I tried to admit him to discover that he no longer had a valid clinic card. This required me to certify him as having a life-threatening illness in order to get him admitted. When I asked about this, he told me that his green card had been lost in a fire and that he could no longer prove his legal residency. For the first time I doubted something he told me. I had heard the "lost in a fire" green card excuse before. It was a standard, unimaginative excuse for many who did not have the Immigration and Naturalization Service's (INS) permission to be here. I didn't blame him for trying to stay here longer than he was legally allowed, but I couldn't understand why he couldn't tell me about it honestly.

After that admission I had to either pull strings or sneak Régis into my office in order to see him. Obtaining even routine laboratory tests was next to impossible without a clinic card. Margaret had begun a study using interferon as therapy for patients with AIDS and was giving it to Régis, but again, without a clinic card, it was impossible to monitor the effects.

To make matters worse, he was not rehired for his teaching job. He claimed that his employers did not know he had AIDS but did not rehire him because he had missed too much time being ill. He took a one-room apartment with a friend and shared the rent. He hoped to live off his savings until he got better or we cured him.

Régis told me he thought his mental powers were deteriorating—that he could not remember or concentrate as well. This frightened him greatly. When I tested him with serial calculations, remembering large numbers forward—and backward, and with proverb interpretation, I could detect no deficiency. Although still meticulously neat, he no longer came to the office in a suit coat but rather in a plain white shirt, open at the neck, with no tie.

At the time we were fortunate to have on staff a compassionate and dedicated social worker named Alina, who always welcomed a new challenge. One day while Régis was sitting in the waiting area, I called her into my office. I asked her if she would be willing to take on what might be a most difficult case. I outlined Régis's illness, his personality, his current difficulties with the INS, and the social problems his illness was causing him. I told her I was beginning to believe there were no solutions to his problems but that he was one of the most extraordinary people I had ever met. At first she was reticent. I knew she had recently been hurt by letting herself get emotionally involved with patients and their frequently insolvable problems. She told me as much and said she would think about it and let me know by the following day. I didn't push. Perhaps she was having a bad day. Before I had finished with Régis on that visit though, she knocked on my door and told me she would accept my offer and try to help.

When I introduced Régis to Alina, his charm and smile returned for the first time in a good while. She talked with him for an hour. I did not see either of them that day after the interview. The following morning I asked her about her thoughts. She said, "You were right, Art. He is an incredible person."

"The man is a saint," I replied. "There certainly doesn't seem to be any justice in what life has thrown his way." Alina agreed.

She had advised Régis that the only way to begin to help himself was first to prove the validity of his residency in this country. Once that was accomplished, she could arrange for him to get temporary financial assistance and a clinic card, which would allow him to receive medical care. He said that if he went to the INS, he would surely be arrested and deported. Furthermore, if he was sent back to Haiti he would most certainly die. She persisted that this was the only way we would be able to help him. He said that he would think about it.

I saw him weekly, mostly to test his mental status and reassure him that he was not deteriorating. He seemed to depend on these visits for his sanity. I got the impression that the people in our office were the only people in the world with whom he shared his secret. He would always have a few words with Amal while he waited and then would visit at length with Alina when I was through with him. I was grateful that she had relieved me of major responsibility for his emotional support. He was so articulate. To hear his thoughts and fears and not be able to do anything about them left me drained.

The more we pushed Régis to go to the INS the more he resisted. He seemed more afraid of returning to Haiti than of being destitute and dying in this country. I wondered if he was in trouble there or if there was something else he was hiding from. Alina thought the explanation was simple: If he went back to Haiti now, his whole life would be a failure. He was not ready to face that prospect. He chose instead to try to survive from week to week, hoping, with some encouragement from Amal, for a miracle.

Revelations

1984 FOR SEVERAL WEEKS MY MOOD was clouded by a story I saw on the evening news. The bodies of 32 Haitians had washed up on a beach just north of Miami. Their homemade sailboat had broken up at night during a storm only 200 yards from shore. The images on the screen were both horrible and fascinating. The sailboat was still recognizable, half-buried in the sand, with waves breaking over it. With honesty rarely shown by television news, the dead Haitian men and women were shown strewn along the beach almost as if enjoying a holiday. Most were nude and in death beautiful. Their bodies were young and muscular, and their skin was textured by sand and beads of water sparkling under a brilliant sun. They must have died shortly before the cameras arrived, for rigor mortis had not yet set in, and they had not yet become bloated. Each wave would wash over them and then ebb, moving them just enough to make them appear to be still alive. Most were face down, looking perfectly at peace. The next day the television station that carried the story apologized for the graphic footage the night before; they claimed it was a late-breaking story.

A few who survived the ordeal had been taken to the Krome Avenue camp. They told how they had signed up with smugglers for

$200 per person and how 40 people had crowded into the small open boat and set sail up the windward passage. Two hundred yards from shore, just before dawn, a squall capsized the boat. Most were just too weak from the journey to make it the remaining distance. The survivors were not even allowed to attend their shipmates' funerals.

Amal and I discussed the tragedy during our ride home. While our rides to and from work were frequently taken up by theoretical discussions of theology and philosophy, this time I spoke directly to the point.

"I'm angry with my government for forcing disasters like this to happen and ashamed to be an American," I told her. "Everyone gives lip service to love and freedom and uses them to sell everything from hamburgers to insurance. But when you cut away all the crap, nobody really cares about these people. How many others do you suppose have been swallowed up by the ocean without a trace? At least these 32 were lucky enough to make the news for their efforts. Altruism is a luxury you have to be able to afford to dabble in. Look at poor Régis. All he wants to do is complete his education and then go back and fill and pull teeth in his country. Now he is forced to go begging." My voice was raised, and I was talking to her as if the boat tragedy were her fault. I knew I was being unfair to her: since she was Egyptian, she was not to be blamed for our national hypocrisy. What I was really angry about was her persistent optimism that all things would work out for the good.

When we had reviewed our data and reinterviewed our surviving patients in response to our critics, we discovered that indeed a small minority of the Haitian men had had homosexual experiences. This fact was seized on by our detractors, who ignored the fact that in most of our men, and all of our women, there were still no reports of homosexuality. New theories appeared in the newspapers every day: AIDS started in New York, transmitted by gays vacationing in Haiti to destitute Haitian men forced to prostitute themselves; conversely, AIDS started in Haiti as a tropical disease and then was transmitted to vacationing gays by the same mechanism; AIDS started in Africa, brought to the Caribbean by Cuban troops returning from Angola and then was transmitted to the gay community. Again, there were no hard facts to support any of these theories.

There were five patients with AIDS on my ward team at that time (early 1984), including Herminio, the first gay man admitted to my service who survived long enough to be followed by me in my office. Herminio probably lived his entire life without anyone, other than his lovers, knowing that he was gay. He was 42 when he was admitted with pneumocystis pneumonia. His hair was gray and conservatively cut. He had worked as a doorman at a hotel on the beach, but he could have easily passed for a teacher or bank teller. He was extremely frightened by his illness. Even when he was clearly getting better he would ask us if it was time to call the florist—a Cuban cultural expression meaning, "Am I going to die soon?" He lived alone with his mother, and he asked us not to tell her what the matter with him was. After his discharge he did well for a while, a fact he attributed to my personal attention to his case.

Fanese was perhaps the most pathetic patient of the first wave. She had a viral infection, for which we had no treatment, that slowly took over her entire body. While in the final stages of deterioration she gave birth to a baby girl, who was also discovered to have the syndrome. The father was unknown. No agency would accept the baby for care, knowing its diagnosis. By the time I took over the inpatient service that particular month, the decision had been made that we had done everything we could for Fanese and that she was failing despite our best efforts. We would try to make her comfortable, but there would be no more "heroics." She was alert but profoundly depressed and too weak to get out of bed. She remained in this condition for two weeks before she became stuporous and then unconscious. She died a week later. In an act of remarkable compassion, one of the nurses working in the hospital who was aware of Fanese's plight adopted her child.

For one reason or another, most of the original study group had stopped seeing patients with AIDS. Only Margaret, Gordon, and I continued. Because of my other responsibilities and commitments, I followed the fewest. Many of the gay men Margaret followed had been coming to the office long enough now that their deterioration was painfully apparent. Men who originally appeared healthy were now coming in wheelchairs or with pillows because they were so wasted that sitting on a plastic seat too long without a cushion of flesh was unbearable. Many now looked like old men, with sparse white hair, wrinkled skin, and masklike faces. The secretaries in my office were genuinely moved by their suffering. They knew them all by their first names. Frequently they would be admitted to the hospital and then be absent from the office for several weeks. When they returned, they invariably showed signs of deterioration. When the secretaries saw them in this worsened condition, they were shocked and frequently hid in one of our offices and cried. When one of them would pass away, no one would commiserate with Margaret more than they would. Margaret took each death personally and after each would mourn for several days. During these times it was difficult for anyone to reason with her.

Meanwhile, Alina was talking to Régis almost every day. She had managed to get some support for his rent from Catholic Charities, but his situation was rapidly becoming desperate. He admitted that he had not been entirely honest with us about his immigration status. Apparently he had come to this country three years before on a student visa, which had expired.

We sent two letters signed by Margaret, Alina, and me to the INS explaining Régis's situation and asking that he be allowed to stay in this country for humanitarian reasons. Neither letter was ever formally answered, and daily phone calls to the responsible bureaucrats were not returned. Amal prayed with Régis while he waited to see me. In the privacy of the examining room, Régis told me that he appreciated all that we had done for him, but that he felt the situation was becoming hopeless. He wondered why God had forsaken him. I reminded him that I had promised him during his first admission that I would tell him when I thought the situation was hopeless and that I didn't think we had reached that point yet. Although his financial and social problems seemed insurmountable, his T cells were still holding their own. I agreed with him that returning to Haiti would be disastrous, for I was sure there were no resources there to treat his illness. I told him that he had to trust what Alina had said. If there was to be any hope of him remaining in this country he had to go to the INS office and straighten out his immigration status. Then Margaret and I could continue to treat him and Alina could arrange for some financial assistance.

Régis complained of losing vision in his left eye. I tested his vision with our eye chart and found that his vision was still 20/20 in both eyes. His eye examination was normal. He was almost as frightened of blindness as he was of losing his mental powers. I tried to reassure him that his vision tested normally, but he was adamant that his eyesight was failing.

I came to my office one morning and found Alina fighting back tears. My first thought was that she was going to tell me that Régis had died. Instead, she told me the following:

"I'm so angry at myself, Art, for not listening to him," she said, composing herself. "It was exactly what Régis said would happen. For months I have been telling him that he had to go to the INS office and straighten out his immigration status. He told me that if he went there they would arrest him. I assured him that it wouldn't happen, and when he finally took my advice, that's exactly what happened.

They called me this morning from the emergency room. There are two versions of what happened. Régis claims that he went to the INS office, and after waiting for a long time in line he started not to feel well and asked if he might be moved ahead in line. When the others in line found out he was a Haitian and didn't feel well, someone yelled, 'Hey, you've got AIDS!' Suddenly they were beating him and kicking him until he fell to the ground. Guards came and arrested him and took him to Ward D [our prison medical ward inside the emergency room]. The charges against him were dropped when they found out he had AIDS. Evidently, they didn't want to have to touch him after that.

The story that the police told the doctors in Ward D was that Régis had tried to cut in line and then began fighting with the others in line and spat at them and told them he had AIDS. He was arrested for disturbing the peace. By the time I got to the emergency room he had disappeared.

I tried to convince Alina that she could not blame herself for the suffering Régis had gone through. We had all heard stories about what it was like at the INS, I told her, but given his situation we had no other choice than to recommend what we recommended.

She wondered out loud why there were two versions of the story. Was there any truth to the police's version of what happened? I told her about the pictures in Régis's album, the ones where he was photographed with the missionaries and the children. Was the person we knew capable of initiating that sort of violence? I had no difficulty deciding which version of the story I would believe.

Régis showed up unexpectedly at our office three days later. Since fleeing the emergency room, he had been living on the streets. He still had bruises on his face and arms from the beating he had received. His clothes were wrinkled and dirty. I told him how sorry I was for all he had gone through. He was very concerned that his vision was now deteriorating rapidly. When I examined him I found spots on the retina of his left eye, which indicated a viral infection. He had been right about his vision all along. It had just been too early for me to pick up on examination. I spoke with Margaret. There was a possibility that a new antiviral agent might slow or reverse the progression of his blindness. I admitted him to my ward team.

Riding home that evening, Amal and I were somber. I asked her if she believed in the second coming. "As sure as my next breath." she responded.

"And what is prophesied about the Second Coming?" I asked.

"The Book of Revelations says that He will come in a time and in a manner that is least expected." She lowered her eyes in reverence as she always did when she quoted or paraphrased the Bible.

"And what if I told you that Jesus Christ came for his Second Coming as a young Haitian man with AIDS? And what if the entire world missed it? Régis, art thou a king?"

"Don't speak such blasphemy!" she gasped.

"What blasphemy? He may not be crucified but he has certainly been beaten and spat on. Actually, he is being crucified. It's just that it's playing out over three years rather than three hours. If he is not dying for his own sins, he must be dying for ours. Another innocent on the altar. Another virgin in the volcano. You want blasphemy? Not only did you miss the Second Coming, but probably a third, fourth, over a million comings. He comes every time an innocent suffers unto death."

She was silent for a long time. I'm sure she thought that if she provoked me further I would lose my soul forever. Just before I let her off at her home she asked if there were anything she could do. I surprised her by saying, "When he is ready to go home from the hospital, take him home with you."

"Take him home with me?"

"Well, isn't it written, 'and I was naked and you clothed me, and I was starving and you gave me food, and I was homeless and you sheltered me?' You're missing your big chance!"

"I would take him home with me if I had a place of my own."

"You have your own room."

The thought of this Egyptian woman taking a Haitian man home to our neighborhood and nursing him back to health was beautiful enough to intrigue me. But I knew that I was kidding myself. And she knew I had won our philosophical war. Such things were just not done. She said nothing further as she got out of the car and entered her house.

Sainthood

I RAN INTO MY RESIDENT on my way to rounds the next day. I told him I had admitted one of the patients with AIDS whom I was following. "Which group of the '4-H' club does he belong to?" he asked. According to the house staff, the "4-H club" stood for the four groups at risk for AIDS—homosexuals, Haitians, hemophiliacs, and heroin users. I admonished him not to be so cynical and told him that this patient was special to me. I told him how Régis had worked as a dentist in Haiti, how that was probably how he got his disease, and how we ended up being "blood brothers" as a result of the marmoset experiment. I then told him the circumstances of Régis's admission—how the blindness had been developing and how he had been beaten at the Immigration and Naturalization Service Office.

Régis had spent the night in the emergency room because there were no beds available on the medical service. I had to admit him through the emergency room and declare him life-threateningly ill in order to get him into the hospital, since he could not prove that he was a legal resident of Dade County, Florida. After spending most of the night on a stretcher in the emergency room, Régis was brought to his room on North Wing II. This was the oldest part of the hospital still being used for patient care. It was depressing just to be there. It was darker than the rest of the hospital, the rooms were smaller, and the paint was flaking off the walls, which were colored a hideous green. Régis was deeply depressed and was sure he was not going to leave the hospital. When I approached from the left side, he could not see me. I asked him how he was. He told me he was tired of being constantly asked if he was a homosexual. Evidently, his treatment in the emergency room had not been kind. Something about his mannerisms made me worry he was going to snap under the strain. His hands trembled when he reached for something and when talking to me he would play with his bed clothes or sheets between his fingers and avoid eye contact in a manner that was new for him. I knew how much the blindness frightened him. I asked the house staff to call Jeanette, our Haitian-American psychiatry liaison, to ask if she would see Régis for emotional support. Later that day I saw Jeanette at a distance down one of the corridors and waved to her. I asked if she had seen Régis yet.

"Oh, I certainly have."

"What did you think?" expecting an outflow of empathy and amazement.

"Unbelievable."

"I know."

"He is a phenomenon."

"Yes, he is," I said, but did not really follow her.

"I mean, I have heard about cases like this, but I've never actually seen one. He really is extraordinary."

"What do you mean?"

"I really don't know quite what the word for it is. It's not quite 'social climber,' but he is acting so much out of class. He comes from poor rural Haiti, you know. He would only talk to me in French. He would not talk in Creole, in spite of the fact that I speak Creole fluently. And this business about dentistry by correspondence. He is really trying to make it."

"Jeanette, I asked you to see him because he's depressed and suffering. What do you care if he didn't go to Harvard Dental School?"

"Because I think he is lying to you. I think you've been had."

"What do you mean 'had'?"

"I think he is gay."

"Gay! You think that's how he became infected don't you? I've asked him about this over and over. He's a dentist. He leads the life of a saint! The man's a saint!"

"How many saints do you suppose were really gay?"

"Do you have any proof?"

"No, nothing specific, just a feeling. With these sociopathic personalities you can't take anything they tell you at face value."

"Sociopathic!" I screamed. "The man spent five years of his life pulling teeth in the backwoods of Haiti. I've seen the pictures."

"You poor bleeding heart. You idealize everyone. Besides, I've heard of this character from the Haitian community. He is aloof. He takes a superior attitude."

"Jeanette, did you do anything to help him?"

"Oh, yes. I was very supportive of him in his struggle against his illness, but I just wanted you to know he's not who you thought he was. I'll continue to visit him daily. Perhaps I can find someone in the community who will take him in."

I left for my office upset but realizing that Jeanette's assessment had to be accounted for. After all she could talk with him in his own language and I could not. She knew the culture and I did not. Worse yet, she had nailed me with a variation of my old college canteen theory. Alina was waiting for me when I got to my office.

"Did you talk with Jeanette?"

"Yes, I did. Art, I can't stand it. I can't stand being taken advantage of and lied to."

"How do you know you were lied to?"

"She says he's gay, Art. I believed every word he told me."

"He's one of the few patients with a negative test for syphilis," I responded weakly.

"She's from his own culture."

"Yes, but I've known him for over two years. He's just a gentle, intelligent man. I've seen the pictures of him performing dental extractions. It was the tainted blood of one of those extractions that infected him. She is a human being just like us and just as likely to be wrong."

"I can't bring myself to go and visit him."

"Because he might be gay or because he might have lied?"

"No. If he were gay I could understand why he might have to lie."

"Look, let's suppose she's right, and he led us down the garden path for all these months. He is still sick, suffering, and destitute. Does it really make any difference? Why do we keep trying to make AIDS someone's fault?"

Although Régis received seven days of a new intravenous antiviral, his vision did not improve. My resident was on my case. He told me that my "blood brother" was trying to use the hospital as a hotel and that I had to help get him discharged. I talked with Régis privately and told him that there was little else I could do for him. I asked him to investigate all of his resources and see if he couldn't find someone he could go home with. He told me he had no one. Reluctantly I told him that perhaps it was time to think about returning to his family in Haiti. At least they could give him food and shelter. He looked at me in despair and said, "I came here to succeed. If I go back, I go back in disgrace. I surely will die." The next day on rounds Régis told me he had found somebody to go home with. I was surprised by this and didn't quite believe it. I told him I wanted to see him in my office in a week. The housestaff had already written his discharge order.

Güinen

DURING THE FOLLOWING WEEK I examined my own ambivalence toward Régis. I weighed the possibility that he was gay and had lied to us over and over again. There was strong evidence on both sides of the argument. Jeanette was a good psychiatrist, and his homosexuality would explain much of his mystery. It played to my personal first principle of psychiatry: Things are never what they seem. On the other hand, I usually know when someone's lying to me. I was willing to grant that he was proud and striving to achieve more than he was born into. This just made him more inherently noble. He could be forgiven the sin of aloofness, but there had never been anything but honesty in his relationship with me. When I had doubted his assessment of what would happen if he went to the Immigration and Naturalization Service, I had been wrong. When I had doubted his blindness, I had been wrong. If I doubted him now, I was at least as likely to be wrong again.

Either way really made no difference. Operationally things were the same. He did not deserve persecution. The death sentence had already been passed. There was no way I was ever going to know for sure. And why did he need to have gotten AIDS from being a dentist and not from being gay for me to like him so much? This last thought, too unsettling for me to dwell on at length on any single occasion, haunted me for several years. Perhaps I was not as free of fears, phobias, or prejudices as I wanted to think I was.

For the most part I was depressed that Régis's last hospital admission had been such a failure. He left with even worse vision than he came in with. His long-term prospects for staying in the United States or surviving until we had some sort of treatment were as remote as ever. And a seed of doubt about his honesty had been planted in the minds of the only people in this country who cared about him.

Alina also had resolved her ambivalence. Although we never discussed it, I'm sure her thought process was similar to mine. She became worried when we did not hear from Régis. It was possible that he was living on the streets again. The friend he went home with might not have existed. We were both concerned that the breach of trust from the last admission might keep him from coming back. Why did we ever question his honesty? A week passed.

He came to the office only one more time. I was not there, but he spoke with Alina. He told her he had decided to return to Haiti. Alina made arrangements with the Catholic Charities to pay for his airfare back to Haiti and put him up in a hotel for a few nights. He would be leaving in a few days. As she told me of the plans for his departure her eyes welled up with tears, and I had a terrible lump in my throat.

"He told me that he wanted to thank you for all that you did for him. He said that perhaps if he gets better he will return and complete his education." I told her that I had no delusions. Once he got on the airplane neither of us would ever see him again. She said she was meeting him at the airport and asked if I would like to come.

I looked at my appointment book. Since I was on the hospital wards in the morning and had private patients scheduled in the afternoon, it really wasn't possible. Yet I couldn't help feeling relieved telling her I could not go. It was a precedent I did not want to establish. I cared about Régis. I felt a loss in his leaving and a failure in not being able to do more for him, but I had to keep my distance. It was the only thing protecting me from the pain I saw Margaret experience with each death. Still, I regretted that I did not have one more chance to see him.

Alina met Régis at the airport as she had promised. She sat with him and waited until the time of departure. He was wearing his three-piece suit again. Her description of his departure sounded like one of those newspaper accounts of a prisoner going to the electric chair. I was even more relieved that I hadn't gone with her. Evidently somewhere during the days since his discharge Régis had regained his composure. She told me he was calm and seemed resigned. Once again, he thanked her for all she had done for him and asked her to thank me one more time. He pressed her hand, smiled, and then boarded the airplane alone. I knew it would be only a brief stop in Haiti before La Sirène escorted him to Guinen. As I later learned, in Voodoo cosmology the souls of the dead are guided by a mermaid, La Sirène, under the sea back to Guinen, their ancestral homeland, the Haitian equivalent of paradise.

Danse Macabre

1985 DAN AND I TOOK MARGARET aside and talked with her about starting a special clinic outside our office for patients with AIDS. Both staff and private patients could be seen there by the faculty. This would facilitate her research and also provide a service to the house staff, for AIDS patients were usually too ill and too complicated to be followed in the general medical clinic by the residents. Gordon and I agreed to continue to help her. Euphemistically, we called it the Special Immunology Clinic.

The danse macabre continued. Previlus died while I was away on vacation. He came into the emergency room dehydrated from diarrhea, as he had done so often in the past. Then he unexpectedly had a cardiac arrest. He was resuscitated and spent his last few days in the intensive care unit before he gave out completely.

Of the original group of patients only Theophile was still with us. For four years he did remarkably well. I asked Gordon what the secret of his success was. "I don't know, maybe Voodoo," he said, half jokingly. Soon thereafter, however, Theophile took a turn for the worse and was admitted to the hospital at the same time that Previlus died.

Other patients played out their illness shortly after Régis left.

Julienne suffered before she died probably more than anyone. She had AIDS on top of diabetes and was afflicted with the worst case of oral thrush and genital herpes one could imagine. The herpes blisters were so extensive and weeped so much she could not sit down without staining her bed clothes. During her last few months I had her in the hospital five times—twice for toxoplasmosis, from which she made a dramatic recovery, twice because the thrush was so bad she could not eat, and once for tuberculosis. Through all this suffering she never failed to smile when she saw me or thank me as I left. She died of a disseminated virus.

Herminio did well for a while but then began to decline. He started to age visibly between visits like so many of the patients Margaret used to follow in the office. We diagnosed tuberculosis, and I hoped he would stabilize with treatment. He still had not told his mother what he had. After each visit he apologized for the inconvenience he had caused me. Each time I told him it was not his fault. He looked at me directly and answered, "Yes, it is."

The Special Immunology Clinic was held every Thursday afternoon. The waiting room was filled with patients in all stages of the disease and their lovers and families. The clinic staff showed varying degrees of apprehension. One secretary wore rubber gloves to run the appointment computer. She didn't even work near the patients. On the other hand, although concerned about contagion, the nurses in the clinic demonstrated an extremely professional attitude. They took the temperature, pulse, and weight of the patients before they were seen by the doctor. Even in this high-technology era, these simple measurements were the best way to separate who was doing well from who was doing poorly. They helped us draw blood and cleaned up after the patients. After the clinic was over, they were instructed by the clinic administrators to wipe down the examining tables and furniture with disinfectant and alcohol.

In addition to seeing increasing numbers of patients with the real disease, we began to see more and more of the "worried well." The newspapers ran stories about the spread of the epidemic among drug users. Shortly after these stories appeared, I began to see one or two patients per week who had dabbled with drugs and were now convinced they were dying. Their symptoms were, for the most part trivial—a sore throat that took longer than a week to get better, a bruise that was slow to heal, or a bump on the scalp that had not previously been noticed. Their anxiety was visible from across the waiting room. Ironically, the patients who actually had the disease were much calmer.

About 25 patients were now coming to the Special Immunology Clinic each week. Of these about 15 actually had AIDS and were being followed by Margaret, Gordon, or myself. Another five were "worried well." The remaining group did not have AIDS but had symptoms or signs which suggested that something was not quite right with their immune systems. At the time we did not know what to make of those patients.

Brian was 30, muscular, with blond hair and a great tan. He came to the clinic wearing a cut-off sweatshirt and shorts, which exposed part of his rear and sneakers with no socks. He had a lover who had died of AIDS. He lived in Fort Lauderdale and worked as a waiter. Although outwardly at ease, I felt mildly uncomfortable taking care of the increasing number of gays who were coming into my practice. Although Brian suffered some weight loss and diarrhea, he was still healthy enough to be flirtatious. Or was he just having fun with my discomfort?

"You're kind of cute. I was expecting a lady doctor."

"Margaret and I work together. I'm kind of a silent partner. I don't get much publicity."

"Oh, the Gary Cooper type? Well, that's okay by me."

He then launched into a monologue of the difficulties in getting an appointment, taking time off from work to come down from Fort Lauderdale, having to pay his fee in advance and so forth. Another man had come into the examining room with Brian. He said nothing but seemed deeply concerned and watched my every move with great interest.

In addition to his symptoms, Brian had impressively swollen lymph nodes. I recommended a lymph node biopsy. We had already identified others who seemed to fit into a category similar to Brian's, with swollen lymph nodes, fever, weight loss, perhaps diarrhea or thrush, but no overwhelming infections. They frequently had a relationship with someone else who had the full-blown syndrome.

Brian missed his appointment the following month. A week later I got a phone call. He had not kept his appointment because the lymph node biopsy I had scheduled had not been performed. The surgeon had been tied up with emergencies. I had to call the surgeon, rearrange this procedure, and reschedule Brian back to see me in the clinic. The lymph node biopsy proved inconclusive, but his symptoms had progressed enough to erode his bravura and cynicism.

"Hi, doc. Glad to see yah. I know why I'm here today, but God only knows why you are. I don't know how you can stand dealing with a problem like this. You gotta help me, Doc. I'm not ready to die at 30."

I explained to him that he didn't have the complete picture of AIDS and that we needed to follow him to see if he improved or deteriorated. Margaret had a protocol using a drug that stimulated the immune system. I got Brian enrolled in it.

I told the story of Brian to my wife at supper. As a general rule I didn't talk about AIDS or the Special Immunology Clinic. She knew I worked there every Thursday. Each Thursday when I came home from work she asked me if I had washed my hands. Although she claimed she was being funny, I worried that she was serious. She told our friends who knew I was working with AIDS patients that, if I ever came down with the disease, I'd be "out the door" in a minute.

To say my wife and I are very different is a gross understatement. Our two daughters picked up on this early on. When they were teenagers, they would introduce us to their friends with, "Meet our parents, yin and yang." We came from very different backgrounds. I grew up Catholic and poor. Janet is the middle child of Jewish parents. Her father, brother, and sister are all physicians. What's kept us together through the years are the shared values, clearly inculcated by my Italian mother and her Jewish mother, of commitment to and primacy of family.

In retrospect, the AIDS epidemic was, in Janet's eyes, more of a threat to our relationship than infidelity. Remember, in the early 1980s we still didn't know very much about the disease. What if I died and left her children fatherless, or what if I somehow passed it on to her? More importantly, she probably felt she was losing me to my work. AIDS, in effect, became the "other woman." In truth, the epidemic was consuming large chunks of my life. I was becoming a workaholic, with an 80-hour workweek and two to three hours nightly on homework, writing grants and articles. To me it was a question of balancing two commitments—one to my family and one to my profession. To Janet's way of thinking, there could only be one commitment—to our family.

Our children, Adrienne and Suzanne, of course, took all of this in. Fortunately, I was aware enough of what was happening to somehow time manage my life. I made a commitment to be there for my children. Our daughters, in turn, were always wiser than their years. Not only did they understand their parents' differences, they somehow managed to synthesize the best of both of us. In fact, if the AIDS epidemic drove a wedge between my wife and me, it also forged a bond between me and my daughters. It opened their eyes to responsibilities beyond family.

My older daughter, Adrienne, was 7 in 1983. Before the acronym AIDS had been coined, we had been calling her "Aids" as an affectionate abbreviation. One day she came to us and said she didn't want to be called "Aids" anymore. She had seen on the television that it was a horrible disease and people were dying from it, and she didn't want it as a nickname any longer.

Mother and Child

ANNIE BARELY LOOKED HUMAN when I first saw her. She was admitted with an allergic reaction to one of her medicines. Bactrim had made her skin slough off almost entirely. The tissues underneath her skin weeped fluid and pus. The drug she was given for her pneumonia, the alternative to Bactrim, dropped her blood sugar, which led to seizures. She lay in her hospital bed intubated, unconscious, and twitching continuously. I thought to myself, "We'll never save this one."

Over the next week she slowly improved. When her seizures were controlled with medication, she woke up. She was extubated and breathing on her own the following day, and her skin started to heal. At one point the new skin was a considerably different color from the remaining crusted old skin. She had a marbleized appearance, and I was worried she would be left looking grotesque. By the time she was ready to be discharged, though, she had her normal coffee-brown skin back again. Her swollen amorphous face had turned back into that of a young woman. She was short, neither fat nor thin, animated, and inquisitive.

I saw her name on my clinic appointment list two weeks later. When I came into the examining room, I was surprised to see a four-month-old baby in a stroller next to her. She asked if I minded if she heated up the baby's bottle in the sink while we talked. She was very concerned and attentive toward her child. Several times as we talked the baby cried and she went over to him and tried to get him to take a pacifier.

She was 27. She had used drugs heavily as a teenager. Now she was married to an auto mechanic. She hoped to return to school to get her high school equivalency diploma. Four months after her child was born she was admitted to the hospital with pneumocystis pneumonia. We had already seen some children with AIDS, but this baby boy seemed healthy enough, though, husky and laughing robustly.

In the middle of a sentence, as we were talking, Annie shrieked and jumped off the examining table. She ran to the sink, where she had put her baby's bottle to warm up. The bottle had blocked the drain and the sink was overflowing with hot water. A large puddle was soaking into the rug underneath the sink. She couldn't get the bottle out of the drain because the water was so hot. Her shriek had frightened her child, who was positioned between the sink and me. He started to howl. Since she had put her examining gown on backward, it opened in the front as she jumped off the examining table, and her breasts jumped up and down furiously as she bolted from the examining table to the sink. With no regard for her own modesty, she tried unsuccessfully to shut the water off in the sink and remove the bottle with one hand and give the baby his pacifier with the other. Finally, I shut off the water while she pulled her baby from his stroller and nestled him between her breasts. There, between her breasts, the pacifier, and the bottle, he eventually returned to his former state of contentment. The nurses in the clinic had heard the screaming and were pounding on the door. "Just a little accident. Nothing to worry about," I told them and I found myself chuckling despite myself.

Annie's child made her a "research interest" to the faculty studying how children got AIDS. She asked me if it were possible that her baby and her husband might have the disease also. I told her it was a possibility. The baby barely whimpered when the blood sample was taken. Annie comforted him with a lullaby that was curiously twisted: "Hush baby, don't you cry, momma's gonna be all right."

The numbers of AIDS patients I was caring for and the overwhelming nature of their problems were beginning to numb me. This was particularly true for my Haitian patients, in whom I saw the same vicious cycle repeated again and again. AIDS led to loss of work, which led to enormous social and economic problems and ultimately to death. Although I could talk to them only through a translator, and although I could do precious little to treat their illness or help with their social problems, they continued to smile and thank me after each visit. In terms of sheer numbers, however, the Haitians were decreasing while the numbers of gays and drug users were exploding. It seemed the epidemic was evolving out of control. Every time we figured out a piece of the puzzle it changed into a whole new puzzle.

Time passed week by week, and the number of patients I cared for continued to grow. I can no longer remember the details of each case. Usually, when someone missed an appointment, it meant they had been hospitalized or, worse, had died. Each Thursday some of the patients who came to the clinic were sick enough to require immediate hospital admission. The housestaff began to dread being on call on Thursdays.

AIDS had intruded into my daily life more than I could have imagined. The phone rang and awoke me from sleep all too frequently, with calls from sick or anxious patients. Real and imagined problems surfaced daily. Even during breakfast, as I sipped my coffee and read the newspaper, I recognized an alarming number of names in the obituaries

Meanwhile, Annie was in the hospital three times with pneumonia. Between admissions she suffered with painful swallowing and existed on a liquid nutrition supplement. Unlike many of the patients, her strength did not erode after each acute illness. When I saw her after she was discharged from the hospital, she would usually come into the clinic as saucy as ever. I breathed a silent sigh of relief each time I saw that her child continued to grow and stay healthy. One day she asked me straight out, "Am I going to die?"

"So far no one has lived longer than four years. We can only hope that in the near future we'll find a treatment or a cure."

"It was that damned dope!" she sobbed quietly. "Everyone was doing it. We just thought of it like candy."

She cried uncontrollably for 10 minutes or longer, and I just sat there.

"What about my baby?"

"You mean, if you die?" I tried to dodge the issue of whether her child would get AIDS or not.

"He's just a baby. He'd be all alone in the world without his momma."

"I suggest you make whatever plans and arrangements you can now, while you're still fairly healthy."

"I'm not coming here anymore, Dr. Fournier. I mean I appreciate what you are trying to do for me and all, but why should I pay my husband's hard-earned money when I'm going to die anyway?" With that statement she got off the examining table and began to change without even waiting for me to leave. I made her make a return appointment in case she changed her mind.

At the same time, Herminio was undergoing a slow physical and mental descent into something less than human. He seemed to age a decade between each monthly visit to the clinic. His hair was now totally white. His gums were receding. Large gaps showed between his teeth, so that when he smiled he looked like a jack-o'-lantern. His clothes looked shabbier, and with each visit his belt buckle grew one notch tighter. Soon stains began to show from the back. He let his beard grow, and his hair became matted and greasy.

He refused to go to a nursing home. He continued to live with his mother, to whom he still refused to tell his diagnosis, and a wizened old uncle, who was rapidly looking younger than he did. Toward the end, Herminio was too weak to walk unassisted. They would help him into the clinic by supporting him under each arm. Upon seeing me he would shout with glee, "Dr. Fournier!" His uncle, who spoke only Spanish, would signal to me that he was crazy, by circling a finger next to his ear. While I saw him, he would alternate between laughing and crying. He had an unusual kind of tuberculosis throughout his body, and he was not responding to treatment. At one point he disappeared for several months and I thought for sure he had died. Then he returned to the clinic, looking worse but still surviving.

He was admitted to the hospital for the final time while I was away on a trip. After I returned, Margaret told me she had admitted him but had decided not to continue any of his treatments and to just make him comfortable. I told her I felt this was the right decision and that I would stop by to see him. She said she thought that he would appreciate that. He turned his face from the window as he heard me enter his room. The nurses had shaved him and washed his hair. This restored something of his former appearance. He greeted me with a look of pleasant surprise.

"Dr. Fournier, I didn't know if you would come."

"I was away when you were admitted. Otherwise I would have had you on my own service."

"I think it's time to call the priest and the florist." He managed a faint laugh.

"I'm afraid this time you could be right."

"I'm sorry, Dr. Fournier."

"It's not your fault."

"Yes, it is." With this he turned his face back toward the window.

"Do you want me to say anything to your mother?"

"No, I don't want her to know anything about this."

"Don't you think she may have guessed?"

"I don't care whether she guesses or not; I just don't want her to be told."

We were both quiet for several minutes. I was beginning to feel uncomfortable and tried to think of the right thing to say before leaving. Herminio spoke first. "You are like God to me, Dr. Fournier." He raised his voice and called out to nameless people passing in the corridor: "Dr. Fournier is God!" He grabbed my hand and kissed it. "Whenever I see you, it gives me hope."

"Look at you. Can you seriously believe that? I've been your doctor for the past year and a half, and you've gotten continuously worse."

"No matter. When I see you there is hope."

"We're not giving you medicines anymore. You will be going to a nursing home as soon as they can locate one that will accept you."

I left the room and quickly found a sink to wash my hands. Herminio's kiss had unnerved me. I visited Herminio two more times. The third time I stopped by his room it was occupied by another patient. The social worker on the floor informed me that she had found a place for him and that he had been transferred there earlier that morning. Although he was given a follow-up appointment, he never came. I assumed he did not last long in the nursing home.

Lespwa (Hope)

A YEAR HAD PASSED SINCE Régis left. The legal status of Haitian refugees was still unclear. Although they had been released from the Krome Avenue camp, a bill to grant them legal residency was defeated in Congress. The Coast Guard was intercepting their sailboats in the Windward Passage and returning the passengers to Haiti. Occasionally there were stories on the news of Haitians marooned on desert islands or drowned in the process of "interdiction." No one seemed to know how many made it to Miami or how many were lost at sea. For the most part, the issue of Haitian refugees had been removed from the national consciousness.

I asked Margaret if she had ever heard anything about what happened to Régis. She had contacts on the island, which she had established during her trips there.

"Yes, I did," she said and bit her lower lip. "He just went to his room and closed the door. He refused to see his family or friends. Finally, after several days, when they heard no further noise, they went in and found him dead."

"Are you sure?"

"Yes, I'm sure."

The Special Immunology Clinic continued to meet each Thursday. The waiting room was now "standing room only," and we were thinking of holding the clinic more than one day a week. I finished each session exhausted. Hope, however, was beginning to return. The virus that caused the disease had been isolated. A vaccine was being considered, and the first drugs that we hoped would kill the virus were being tested. I'd begun to tell patients in the early stages of the disease that there might be a treatment or possibly a cure in the near future.

Annie came to the office for me to sign a form. She had just been discharged from the hospital after treatment for her fourth episode of pneumonia. The form she needed to have me sign was necessary in order to place her child in foster care.

"Can you hold on for a few more months? I think we may have a treatment."

"I don't think so. I've lost all my strength. I can't keep anything down."

She looked at me directly. Her face was gaunt, and she was beginning to lose weight. She neither smiled nor cried but stated calmly, "I know I won't be alive in two months." Other patients in the waiting room heard her say this, and suddenly everyone was watching her. She gave me one more look that said simultaneously "Thanks" and "its okay if you can't do anything for me anymore." She folded her form, tucked it in her purse, and left. I never saw her again.

I had seen Lee off and on for two years. Originally, he was one of the "worried well." He had a lover with Kaposi's sarcoma and he would come anxiously with every new bump or freckle. Now he had two weeks of fever and sore throat. His examination showed thrush. He took the news quietly. Then he told me he would take any risk necessary to have a chance to be spared the agony he had seen his lover go through. Although he was white and American, something about Lee reminded me of Régis. He dressed meticulously, as Régis did, and he spoke with a natural eloquence. I excused myself to find Margaret.

She had spoken with someone at the National Institutes of

Health the night before. Dr. Anthony Fauci was looking for volunteers for Phase I studies of an antiviral agent that held the promise of actually killing the virus. Previous drugs had merely stopped its growth. Phase I studies at NIH in Bethesda, Maryland, were designed not for treatment but to measure toxicity. Lee would need to go to NIH and stay there for four weeks to submit to the tests. If the drug proved safe, he could continue it in Miami. He qualified because he was in the early stage of the disease. He happened to have two weeks of vacation coming and could take time off from work.

When we offered Lee the chance to enter the study, he said he needed to talk with his family first. Two hours later he called me back and said he would go. Although I hoped that the drug would live up to its promise, I had lingering doubts. At the very least, I hoped it wouldn't hurt him. Meanwhile, it was Thursday, and there were seven patients waiting for me at the Special Immunology Clinic.

I told Jeanette Margaret's account of Régis's passing. "It's probably not true," she said. "There's a rather infamous gay—very wealthy—with the same last name as your patient Régis. He died recently under similar circumstances. I'm sure that's who her contacts told her about. I told you that Régis was from a poor part of Haiti. Once he returned there, I'm sure he was never heard of. Who'd care? For all you know, he could still be alive, up there in the mountains. It's healthier up there, you know."

"Who are you kidding?" I thought. There's no way he could still be alive. Hope is a funny thing, though. In spite of all I knew about the disease, I hoped against hope that Jeanette was right. At the very least, I hoped Régis hadn't died alone, as Margaret had described it.

Anniversary

1991 THE TENTH ANNIVERSARY OF THE first reported cases of AIDS was approaching. Of course, my hopes for a cure were naive. All of my early patients had been dead for some time. There were few reminders of those days, really. My career took unexpected turns, taking me first away from AIDS patients and then back with a vengeance. Due to a visa problem, Amal returned to Egypt. When she left, she wished me a thousand blessings. Each blessing turned into another patient with AIDS.

AIDS was still making news. The big issue in the press was whether all physicians and dentists should be tested for the virus. This issue was being pushed by an unfortunate young Florida woman who claimed she had contracted AIDS from her dentist. The controversy brought things a little too close to home for many of my colleagues. Most doctors in practice had had enough. They didn't even want to talk about AIDS anymore, much less care for the growing number of patients.

The housestaff quietly fulfilled their responsibilities toward AIDS patients but with no enthusiasm and with occasional hostility. The spectre of contagion haunted my profession. Most physicians wanted all patients tested. Medical students, in particular, seemed torn by the attitudes reflected in the media and what they were taught by their professors. Each year fewer undergraduates were choosing medicine as a career. The most commonly quoted reason: fear of AIDS.

I found an unexpected few moments to speak with Margaret at a meeting in Seattle. We both smiled ironically knowing that we had to travel across the continent in order to have the opportunity to talk to each other. We never saw each other at the medical center those days. Margaret had become an AIDS specialist. No, more than a specialist, an authority. She had conducted the clinical trials that led to the introduction of zidovudine (AZT), the first medicine found to be partially effective against the AIDS virus, in 1987. She addressed the audience of fellow faculty in Seattle with total command of her subject matter.

"Zidovudine prolongs life. DDI and combined chemotherapy are coming, which will further improve survival," she encouraged.

Margaret, however, had paid a price for her fame. The *Miami Herald* had run an exposé on her in the Sunday edition a few months before, alleging she was "on the take" from the drug company that makes AZT, and even attempting to raise scandalous issues about her personal life. I assumed this misplaced medical muckraking was instigated by patients who were frustrated in their attempts to be enrolled in her protocols. I wrote a letter to the editor defending Margaret, which was published, but few others rose to her defense. After I left the Special Immunology Clinic three years previously, she made national news with her investigations on heterosexual transmission among prostitutes in Miami and her clinical trials of AZT. She had six faculty members and a host of support people working under her. The hospital had opened a wing for her patients. Yet AIDS activists pelted her with eggs at an international AIDS conference, accusing her of delaying the arrival of new medicines with her research. She had no personal time. Her telephone was always busy.

In a quiet moment before her talk, she told me that the World Health Organization predicted that by the year 2000 there would be

5 million babies dead from AIDS, 10 million children orphaned, and 15 million dead adults. "What we have is a pandemic. It will be followed by a second pandemic of drug-resistant tuberculosis. We've not seen anything like this since the Black Death of the fourteenth century."

"So if there will be 20 million dead and 50 people infected with the virus for every one with symptoms of the disease, that makes something like 10 billion infected people by the turn of the century," I quickly calculated in my head. These numbers were greater than I could comprehend.

"How many AIDS patients do you think one doctor can care for?" I asked her. "It's exhausting work."

"Five or 600 at best," she replied thoughtfully.

"So we need a city of doctors worldwide full-time to care for all these patients. There is no way the medical profession can make that kind of response."

The image of poor people in Africa, Asia, and, of course, Haiti dying with no medical care overwhelmed me. To me it was clear. AIDS had become a disease of the poor. I wondered if Margaret appreciated the paradox of what we talked about and what she was saying in her lecture. In her lecture she talked about more things we could do for people with AIDS. Newer drugs, better drugs, more expensive drugs, more tests, more physician time, but still no hope for cure. We were going to end up with two kinds of AIDS—for the minority with money, a chronic illness measured over years, with treatments and blood counts, and alternative treatments, protocols, and hope. For the poor majority, insidious weakness, a few hopeless months of wasting, or several hours of suffocation. Even if we gave the medicines out for free, there aren't enough doctors, nurses, hospitals, clinics, and laboratories or, for that matter, enough understanding among people "at risk" to deal with these numbers. George Orwell was right: We all have to die, but it's better to die rich. The poor always suffer more in the process.

My suspicion that AIDS was somehow differentially preying on the poor had been growing during the late 1980s and early 1990s. Actually, that suspicion had been there initially with our Haitian patients. We had looked at income as a risk factor. Our Haitian patients with AIDS were no poorer than those without AIDS. In retrospect, we realized that we had committed a major statistical blunder. The control group should have been a group of rich patients, not other poor people. So for several years, I went against my instincts and tried to find some other link between my Haitian patients and AIDS.

Then, in 1988, Alina—compassionate, caring Alina—engaged me in helping her with one of her growing passions—bringing health care to the homeless. I think it might have been a kind of "payback" for my getting her involved with Régis. At any rate, the social upheavals of the 1980s—the flood of immigrants from Cuba during the Mariel boatlift, many of whom had problems with mental illness, coupled with recession and the impact of crack cocaine on poor communities, had swollen the ranks of Miami's homeless to more than 8,000. They lived in encampments in the city's parks and under the freeways that crisscrossed the city. As I got more involved, I learned that AIDS had become a huge problem among the homeless.

So Margaret and I were on two different planets, universes apart. Margaret lived in the universe of science—numbers, protocols, statistics, and clinical trials. We need these things in order to make progress. That's not to say that Margaret still did not suffer with her patients. I know she did. But science gave her and them hope. But in the homeless clinic I was working in, there was very little science. Mostly, there were a lot of patients with AIDS, and, although I tried, there was precious little I could do.

In my own mind, Tim was the last person in the danse macabre line, the string of seemingly unrelated people—rich, poor, black, white, gay, straight—being led to death by the AIDS devil. During the Black Death, the danse macabre image represented the belief that the plague chose its victims without respect for education, social position, or wealth. For a while, the modern plague in Miami gave most people the same perception. But to me it was clear: The epidemic was changing. In the future the virus would be targeting the people I'd come to know at the clinic—the street people, the drug users, the prostitutes, the poor would become its preferred victims. Tim was, for me, the last great egalitarian death, the last person in the danse macabre line. That's probably because he died so out in the open. A physician dying of AIDS carries a special poignancy. I knew of other physicians who had died of AIDS, but they died secretly. With the first symptoms, they would quietly resign and move back with their families. Other physicians worked for years knowing they carried the disease but told no one. The truth came out only in their obituaries. Tim was different. He went public.

Physician, Heal Thyself

I HAD KNOWN TIM FOR 10 years before he came down with AIDS. He was somewhat younger than me, and had worked for several months as an intern and a resident under me during my first years on the faculty. He helped me care for Régis and the others from the early days and was one of our best residents. He was always compassionate, caring, optimistic, and hard working. I was pleased when he was appointed to the faculty. We collaborated on several projects. He was dedicated to teaching and creative with education. Tim always looked vaguely counterculture: slightly long hair, but thinning on the top, full cheeks, a small "spare tire" that betrayed an interest in the culinary arts. He usually dressed in jeans, a plaid shirt, and a wide flowery tie. Small talk with Tim centered on the clerkship for third-year medical students he coordinated. He was a friend. Not a best friend, but more than an acquaintance. We collaborated professionally and shared academic interests.

One of my secretaries, Anita, told me she had heard a rumor that Tim had told the entire first-year class that he carried the AIDS virus. This happened during a class entitled "Health and Human Values." I never had thought of Tim as "gay" or "at risk" or, for that matter, "straight." He was Tim, a friend, a colleague, a neuter. Anita was well connected to the rumor mill at the medical center. She considered keeping me informed of campus gossip, particularly that related to sex and romance, as an important part of her job. Her opinions on AIDS and homosexuality were complex. She had not entirely escaped a cultural aversion to homosexuality. She once came into my office after a meeting with an openly gay faculty member and, after a long stare, arms crossed and extended foot tapping, declared, "I don't like him. I think he's trying to get into your pants!" As a person who saw most things in sexual terms, to her gay men were subliminally competitors. Yet she had close friends who were gay. She always lent them a sympathetic ear. She felt deeply and personally the suffering of my AIDS patients. One was the brother of a boy she had dated in high school. She wept for him when he came to my office.

Tim and I had been meeting about establishing a teaching program for medical students in Key West. Tim liked spontaneous, unannounced meetings to present me with his creative ideas. The Key West proposal was just such a burst of creativity: Get the students away from the big city, let them see the kinds of problems they'll encounter in a small town, and use the community physicians rather than full-time faculty as their teachers. He presented his ideas with his usual ironic wit and relaxed demeanor. After Tim left, Anita came in, looking despondently at the pictures on my wall.

"Is Tim gay?"

"What makes you ask that?" I answered, looking puzzled. "I don't know. We've never talked about it. But I have no reason to believe so, and even if I did, it wouldn't be any of our business, would it?"

"They say he announced to the whole first-year class in lecture the other day that he had tested positive for the AIDS virus." She moved from looking at my pictures to studying my diplomas.

"Well, that's one of the craziest things I've ever heard. You know how the rumor mill works. He was probably just acting out a role in the course where the students learn how to talk to patients. I wouldn't place any credence in that rumor until you hear it from Tim himself."

Two weeks later Tim and I met again to follow up on the Key West project. At the end of the meeting he matter of factly stated, "You know that I've got AIDS."

"I heard a rumor that you announced that to the freshman class, but frankly I gave no credence to it."

"It's true. I see no reason to hide it. Having a disease is nothing to be ashamed of. Don't you think our students should know that? I just want you to know because I don't know how long I'll be able to continue to work."

"Have you been feeling sick?"

"Yes. For some time now I haven't felt right. At first I thought it was just tuberculosis, something I had picked up on the Jackson Hospital wards. But when the cultures came back atypical tuberculosis, I got tested. They have me on five medicines, which is a pain, but I feel better now. The thing I worry about the most is the dementia. I think I may be getting that. I forget things all the time. I forget where I'm going, or what I'm supposed to be doing, or when my rent is due. I have to write myself notes."

"Sometimes if you worry about something too much, it comes true just from the worrying," I said, attempting reassurance. I silently shuddered when I heard "atypical tuberculosis." Despite Tim's claims to be feeling better on five medicines, I knew atypical tuberculosis was a late complication of AIDS and incurable. He looked too well to be that far advanced. I promised him that I'd help keep him working for as long as he could, by supporting his salary through one of my grants. Tim liked that idea because the grant work would conform to the vagaries of his illness. It required a lot of written material. He could pick up his pen when he felt well and put it down when he felt ill. He had stopped seeing patients when he found out he had AIDS, but there was more than enough teaching and writing to keep him busy. He hoped to devote half of his time to the Key

West project. The other half would be devoted to teaching about AIDS.

Despite the advanced state of his illness when he was first diagnosed as having AIDS, it took two years for Tim to die. His tuberculosis caused fevers, weakness, and poor appetite. The relentless course of his infection was slowed by antibiotics but never reversed. Once, in despair, he stopped taking his medicines altogether but he felt so much worse, with higher fevers and less appetite, that he started back on them shortly thereafter. I found it difficult to avoid clinically assessing how far Tim had declined each time I saw him. The first outward sign of illness was how he trimmed his sideburns. He trimmed them unevenly, leaving a large patch of white skin on the left side of his face. This suggested either a problem in coordination or an inattentiveness to detail. Perhaps he just didn't care about his appearance anymore. There were subtle changes in his hair. It became thinner and straighter. The fullness of his cheeks and waist slowly disappeared. His enthusiasm for work, however, masked many of these physical changes. He was always smiling. We rarely talked about his illness. The Key West project was going well, and we were collaborating on a new course to introduce first-year students to clinical medicine. Some cynics on the faculty were already dubbing this course "The Tim _____ Memorial Clerkship."

Tim conserved his intellectual faculties for teaching and writing. His only complaint was his repeated conviction that he was losing his mental powers. I began to realize that this was, in fact, slowly but surely happening. He would refer to me by other people's names or stop a sentence in the middle of a thought. Strangely, he would usually become aware of these lapses shortly after he committed them and would then apologize with a matter-of-fact disclaimer: "See, I told you I'm losing my mind."

Major and minor lapses in judgment were also present. "I know he is not well, but he is driving us crazy," sighed Anita. "He doesn't follow university procedure. He doesn't bring us back his receipts for his trips to Key West, and he wants to be paid cash in advance. He claims not to believe in credit cards, and if we don't pay him in advance, he's too poor to go." I later learned that the reason why he didn't have credit cards was that he had overcharged and refused to pay his bills. "Let them go after my estate," he was quoted as saying. Yet with me he always seemed in control of himself. He somehow managed to be, if not optimistic, somehow unaffected by the fact that he had AIDS. I had seen this attitude in others, most notably Previlus. In Previlus's case, I attributed it to poor understanding of his disease. Yet here was Tim, for whom the burden of his disease was compounded by the burden of knowledge of his own fate, and he could still think about ideas and things that had nothing to do with dying of AIDS. I could not understand this. If I found out I had AIDS, I would quietly consider suicide. The knowledge of what was ahead would have been too frightening. Didn't Tim have similar thoughts? Being a physician would have made it easy for him. Before his pronouncement, I had heard him once express vaguely Buddhist ideas. Perhaps that was his secret. The cause of all suffering is desire. Freedom from suffering is freedom from desire, even if that includes a desire to keep on living or to die.

We traveled to Key West so he could introduce me to the doctors he had recruited there to teach our students and plan for the health fair that was scheduled in the next few weeks. We had lunch at a restaurant overlooking the Gulf of Mexico. Tim was openly nostalgic about the city, telling me about the best restaurants and the best places to stay. The thought crossed my mind that this might have been where Tim acquired his infection. Living with AIDS was worth it, he seemed to be saying, as long as it included one more visit to Key West. With places like this, life couldn't be so bad, could it? The gulf glistened under the December sun, forcing me to squint as I looked into the shadow of Tim's face. Although I could not make out the details, I could see he was smiling, perhaps dwelling on a fond memory. Between the sun, the water, the perfect temperature, and the clapboard storybook houses, it was hard to believe there was any evil or suffering in the world.

"What's it like to be gay, Tim?" I blurted spontaneously. It seemed like a natural question at the time but was followed by an immediate sense that it was the dumbest question I had ever asked. Had I spoken too loudly? Had others in the restaurant heard me? Anyway, it was out now. Tim had always been so open with me. I did not mean the question pejoratively. I just wanted to understand. I might never have another chance to ask it.

"Forgive me, Tim," I said softly in order to explain, "but when I was growing up, no one admitted to being gay. When I was 13 and my father had to tell me about 'the birds and the bees,' he told me only two things. When I was married, I would 'plant my seed' in my wife and we'd have a baby, and if any other guy ever touched me, to run away as fast as I could. Gays were called 'queers' when I was in high school and college. Even when I was in medical school, homosexuality was taught as a 'perversion.' It was only taken off the psychiatric diagnostic code a few years ago. I'm over 40, and this is the first time I've ever felt comfortable asking someone the question."

"Well, right now that name has a hollow ring to it. I'm not feeling particularly gay right now and haven't for months. 'Resigned' is a more accurate adjective to describe me. There are a lot of us 'resigned' around now, so you better be careful. It might be catching."

"I'm sorry, Tim. I don't want to spoil your day. Am I the first person to ask you that?"

"Before I came out about having AIDS, I don't think people thought of me in sexual terms. I'm kind of nondescript. Besides, I'm a doctor. No, it's really never come up. So, what do you mean? What's it like to be gay and have the whole world stereotype you, or what's it like to have sex with a man?"

"No, not the stereotype. I think I understand that, among other things, I'm left-handed, you know—we southpaws consider ourselves the world's most misunderstood minority. Why do you think you are the way you are?"

"Well, Art, why ask me? If you want to know what it's like to desire a man, why don't you ask your wife what that's like? To be on the receiving end of sex? Again, ask your wife or your girlfriend if you have one. If you think of it that way, those of us who prefer loving men are a majority, and it's me who should be asking you, 'what's it like to be straight?' Stop trying to understand it, Art. Anyway, you've got too much lingering Freudian baggage. Just accept the fact that gayness exists. Always has, always will. Not even AIDS will get rid of us."

We were silent for a while, sipping our beers, me thinking the question had gotten all the answer it deserved, when Tim started talking again.

"Your left-handed analogy is a good one. To me, my 'gayness' is natural. Biologic. It's just that the rest of the world is sexually right-handed, with a few ambidextrous people thrown in for good measure. The interesting thing, though, is when people don't know you're gay. They let slip out some of the most outrageous things."

"How do you suppose my Haitian patients got HIV? Do you believe that stuff about gays from New York on vacation buying the services of poor island boys?"

"I don't know. That may have happened, but I don't think it explains all of it or even most of it. I read a poster presentation at a meeting once that claimed Haitian women allowed anal intercourse during menses as a form of birth control. I never saw that presentation published, thank goodness. It was bad science, with a hefty dose of what I call medical voyeurism. Anyway, I'll bet a lot of straight couples practice anal intercourse, or at least experiment with it just for variety's sake, but aren't honest enough to admit it. That's the problem with any research that has to do with sex. The honesty factor. But I don't think anal intercourse causes AIDS, per se, so I don't think that's how the Haitians got it. And then there's the whole issue of sex during menses. I mean, think about that for a minute as an exposure risk. I think we've just got to realize there's a lot we don't understand. If you threw me on the floor, sat on my chest, and didn't let me up until I told you something, I'd probably say they got it from sex, pure and simple. But maybe it's from medical injections with unsterile needles. Who knows? I'll tell you one thing. That stuff in the *Herald* about blood and Voodoo—that was truly outrageous! Now Régis—he was the exception. He got it from pulling teeth without gloves. Anyway, if you want to find out the answer, my advice is wake up and think out of the box. Don't think like everyone else is thinking. Think left-handed. Think gay. Think Haitian. Otherwise, the virus will always be two steps ahead of you."

Although he was still in the shadows, I could tell that the smile was gone from Tim's face. I suggested we walk to Mallory Square to take in the sunset and the street performers. I noticed an occasional stare from the straight tourists as we left the restaurant and headed down Duval Street.

Meanwhile, Tim kept spending or committing grant funds without accounting for them. My office had to track these expenses surreptitiously. During a one-day visit to Key West he ran up a $100 tab in cab fares.

"Why didn't you rent a car?" I asked in frustration.

"I had to give up my license," he mumbled. "I can't afford the insurance."

He recommended an administrator for the Key West project who did not do a good job. The administrator "retired" to Spain with no notice. The project checkbook was missing for two weeks before it was finally found in a drawer by a secretary. Those two weeks were my worst two weeks on the faculty. Fortunately, there were no funds missing.

These problems with judgment tested my commitment to keep Tim working for as long as possible. "Have you thought about retiring on disability and going home to be with your family?" I asked. Tim's parents had died in an automobile accident when he was young. He was "somewhat estranged" from the other members of his family, except for one sister in Stuart, and he didn't want to be a burden to her. She had young children. Most of his close friends were either already dead or more debilitated than he was. I could see that without his work Tim had little to live for.

Work sustained him. He was hospitalized occasionally to diagnose and treat new infections. Sometimes he would call me from the
hospital if he was concerned he would miss a deadline. At other
times he would just disappear for two or three weeks and then return, as if those two or three weeks had never existed. These absences
were always terminated with spontaneous, unscheduled visits to my
office, during which Tim would try to "sell" me on a new educational
idea. The times in the hospital took chunks out of his body and soul.
Between hospitalizations, his decline was subtle and slow. With each
stay in the hospital, however, he looked more and more like a victim
of a concentration camp.

He gave up his apartment and moved to Genesis House, a home
for homeless AIDS patients. *Why did he do this?* I asked myself. It
was probably more from loneliness and a desire for company than an
inability to care for himself. He was still writing, teaching, and going
to work each day. Or was this the judgment problem, again?

Genesis House didn't work out. The home was not close to public transportation, and the companionship he had hoped to discover
among fellow AIDS patients just did not materialize. He found it
depressing—all these people with nothing to do, waiting to die. He
then became truly homeless. He spent some time with his sister in
Stuart. If he needed to lecture, he would rent a hotel room within
walking distance of the location of his lecture and stay there the
night before and the night after, to conserve his strength. He finally
found another apartment and moved in alone.

There was a small group of students who were devoted to Tim.
They admired his honesty and openness about his illness. They
worked with him in preparation for the health fair in Key West. Tim
had directed the fair for eight years. It was his favorite project. For
the year prior to each fair, the students learned how to do screening
procedures such as Pap smears and breast examinations. On the day
of the fair, people in the lower Keys came for health checkups performed by over 100 medical students. No one on the faculty but

Tim knew the organization and planning necessary to pull off a successful health fair.

As the 1990 fair approached, Tim's decline began to accelerate. His gait became slow and feeble. He added new notches to his belt, which bunched up the top of his pants like a drawstring, similar to what Herminio had done. The students assumed more and more responsibilities. Meetings with Tim were painful. I asked myself, *How much longer can this go on?* Death would have been a blessing six months ago. The administrative part of me wished very much for Tim to retire on disability. Tracking his grant-related expenses had become an office project. The office staff began planning errands around his scheduled visits, not because they were afraid to see him but because they didn't want to cry in his presence.

The health fair went off flawlessly. The students did a great job. Tim summoned what would have been an enormous amount of energy even for a healthy person; finding extra examining gowns, setting up examining rooms, directing students to the screening stations where they were needed the most. Five hundred people came. Three other faculty members, besides me, came down to supervise the students. In general, most faculty members avoided Tim and his projects after he made his announcement. Tim went into seclusion after the fair. We talked on the phone one time before he died. He told me matter of factly that he thought it was time to go on disability. He didn't think he could work anymore. I told him I understood, thanked him for all he had done, and offered to help in any way that I could.

"Art, one more thing." Tim's voice was faltering.

"Sure Tim, anything."

"Promise me you'll keep the health fairs going."

"I promise, Tim. Don't worry."

I heard from Anita two weeks later that Tim had been admitted to a community hospital. Although he had been public about his illness, he was very private about his death. He didn't want to die at the medical center. He didn't want his students to see him dying. His death was mercifully painless. After a few days, he lapsed into a coma and did not wake up.

Tim chose to be cremated and requested that his ashes be spread over the waters near Key West. There was no funeral, but there was a memorial service in the medical school library. Present were his four sisters, their families, several medical students, and a handful of faculty. I was proud to see my office staff standing in the back. Tim's chairman had brief and kind words to say. All of Tim's sisters had tears streaming down their faces. I spoke with one of them afterward. I introduced myself as someone who had worked a lot with Tim. She told me that until that day she had no idea of what Tim actually did or what his work involved. "He never talked about work," she said, "and we just thought he was a regular doctor."

"This is so strange," I thought. I both wanted and didn't want to be there. Death was overdue for Tim. I was relieved for him when he finally died. But the low turnout punctuated how ephemeral his memory would be. He had worked until two weeks before he died. He had made a significant contribution to teaching. I had rarely witnessed such a courageous facing of death. He never complained. Yet now that he was dead, what did it mean? It was as if he expected his peers and his students to learn from his example, but he would not intrude on our consciousness. We had to choose whether we would learn from his experience or continue in our ignorance. He tested us, and most of us failed.

Walking back to the office, I heard Anita mumble, "What a waste." In the past she had used that expression as a put-down, upon learning that someone was gay. "How dare they deprive some needy woman of a husband or lover? How dare their attractiveness be wasted on other men?" I knew, this time, it was AIDS she was indicting, so I let the comment pass, unanswered.

Ninja

1992–1994 OFTEN, THE FIRST THING I saw when I arrived to work at the homeless clinic each morning was Jennifer bathing herself. These daily ablutions took place in a curbside puddle next to the clinic. First she would glance in each direction to make sure that there were no police coming. She didn't care who else watched her ritual; the other homeless people waiting for the kitchen to open for breakfast, the clinic staff arriving for work, or the motorists passing by on Northeast First Avenue. She would peel her tattered dress off over her head. She wore no underwear. Under other circumstances this activity might be seen as enticing or erotic to passers-by. But drugs, schizophrenia, and AIDS had turned Jennifer's 34-year-old body into a parody of sexuality. Her breasts were drooping and narrow. Her buttocks were sunken inward. Her calves were wider than her thighs. The only part of her that was not withered was her pubis, which bulged forward from her lower abdomen like the puff of a dandelion attached to its stem, held upside down. She crouched with both feet in the water and splashed between her legs and under her arms. Next she rubbed muddy water across her chest, scooped some in her hands, and poured it down her back and over her shoulders. Finally, without drying herself, she would put her dress back on and return across the street to the parking meter she lived by. Some days she would repeat her baptism five or six times.

Most homeless people manage to find shelter at night under bridges and in abandoned buildings. Jennifer slept out in the open, under a blanket, on the street corner opposite the clinic. Every night the police chased away the homeless hoping to sleep under the overhang in front of the shelter. Only Jennifer remained nearby. I believe the police were afraid of her. She was so emaciated that, if she pulled her blanket over her head, there was no visible evidence of a human being under the blanket. She looked like an abandoned crumpled blanket on the street corner next to a parking meter. I worried that a truck or bus would misjudge its turn, not care about running over an old blanket, and traumatically end Jennifer's life. From this debris a human being arose each morning and repeated her ritual. If it rained, Jennifer would take her blanket and move under the overhang of the clinic until the rain passed. She was the only one the police never hassled when it rained.

The clinic and the adjacent shelter were Jennifer's protection. A security guard watched over her during the day and most of the night. The Brothers of the Good Shepherd, who ran the shelter, gave her breakfast and dinner. Although the shelter was officially for men only, on the few cold nights each year they would take her into the vacant clinic and allow her to sleep in the waiting room or on an examining table. I wish I had known about the shelter when Régis was alive. Brother Jack and Brother Harry would have protected him, too. They took in anyone, with no judgment passed.

Jennifer had been living on the streets as long as anyone working in the clinic could remember. She had good days and bad days. On her good days, she would wish me a good morning as I walked from the parking lot to the clinic and ask me for quarters. On her bad days, she would wander up and down First Avenue, oblivious to the traffic, answering in obscenities voices only she could hear.

She took no medicines for her schizophrenia or for her AIDS. Although the social workers tried to place her in various shelters, she refused to go voluntarily. There had been a time when she was addicted to crack and supported her habit by selling sex for drugs. Now she couldn't afford her former addiction. She was too ill and too unattractive, and word was out among the street people that "Jennifer's got the 'Ninja'." Ninja is the Miami street name for AIDS. Occasionally there would be a man leaning against her parking meter, usually facing the opposite direction but talking to her. Most of these men had backpacks or bedrolls, suggesting they were new in town. Most of the time, however, she was alone.

She knew she had AIDS but did not understand. Several times since I had started working at the clinic, Jennifer was absent from her corner for prolonged periods. Each time she returned, I asked her where she'd been. "I was in the hospital. I got pneumonia," she would respond nonchalantly. She came to the clinic only when she was in pain. She had sores on her feet that occasionally became infected and needed to be cleaned.

Jennifer was one of 8,000 people living on the streets of Miami at that time. Many had serious substance abuse and mental health issues. These patients were in a medical no-man's-land: deinstitutionalized but not capable of managing their own affairs, lost in their own inner world, and easy prey to pushers and pimps. Treating their medical problems was next to impossible. Others were immigrant women and their children who, upon losing their marginal jobs, were only one paycheck away from living on the streets.

The Camillus Health Concern was located in an older building of vaguely mission architecture. The skyscrapers of downtown Miami were easily visible from the front door. If the building were one story higher, you could see Biscayne Bay and the cruise ships lined up at the port of Miami four blocks to the east. The immediate neighborhood was surrounded by vacant warehouses, abandoned buildings, parking lots, and pawn shops. Litter was everywhere. Rats scurried around the parking lot, ignoring their human neighbors. Two blocks to the west was the Miami Arena, home at the time of our professional basketball and hockey teams. The arena was built there in an attempt to revitalize the area. Many of my patients worked intermittently at the arena, parking cars, setting up staging, and assisting with the concessions. Those who couldn't find these kinds of jobs tried washing the windows of cars passing by or just plain panhandling. The city fathers very much wanted to move the shelter and the clinic away from the arena. No other neighborhood in the city wanted it though.

Every day the food line stretched for two blocks from the kitchen. Every day, when I arrived at work, there were already 15 to 20 patients lingering outside the door, even though the clinic would not officially open for another hour. The windows of the clinic were papered with messages to patients:

> *John Smith, see Joan about your lab tests.*
> *Jorge Gonzalez, speak with Wilfredo about your disability claim.*

Inside, the clinic had a well-used, chaotic look. Boxes of donated medical equipment, supplies, and clothing were piled in corners and corridors. The furnishings of the examining rooms and waiting area had been donated and ranged from the merely old to the archaic. An antique x-ray machine gathered dust in one room of the clinic. We couldn't figure out how to use it. The setting would have been depressing if it were not for the staff of nurses, social workers, and support people who assisted Phil, Jill, Bill, and me in caring for the patients. I'm not making up the rhyming names. In fact, we used to joke about the "Phil, Jill, and Bill" show. Phil looked a lot like Willie Nelson, right down to the ponytail. Professionally, he was an interesting story. A "young Turk" pediatric infectious disease specialist, he had had a falling out with his chair, resigned from the faculty, and entered private practice. He made a lot of money, sold the practice, sailed for two years around the Caribbean, and started to volunteer at the clinic during a refitting layover in port. His volunteering soon became a full-time job—the only job he ever loved, he later claimed. His trademark was a hug. He hugged patients, staff, and medical students—"an equal opportunity hugger," he used to say. The students were soon voting him "best teacher" annually. His gentleness with patients was balanced by his intellect and, when necessary, his acerbic wit.

Jill and Bill were both recent physician assistant graduates. They added the idealism of youth to the experience of Phil and myself. Looking back, these were the halcyon days of my career. My students used to ask me, "Dr. Fournier, why are you smiling all the time?" My quick answer, "Are you kidding? They actually pay me for the joy of caring for these poor people and teaching you!" In our day-to-day work, the faculty and staff all aimed to inspire the medical students by communicating without words that it's okay, even rewarding, to care for those that nobody else cares about. Many of the staff were former street people themselves, like Roscoe, a huge teddy bear of a man who had kicked his drug habit and become a counselor.

The clinic was founded by Joe, one of our residents, who started volunteering at the shelter when he wasn't on call and who was starting to make a name for himself for having the courage to go out under the bridges and expressways, encounter the homeless face to face, and invite them into the shelter and makeshift clinic. Alina volunteered to write a federal grant, which allowed the clinic to expand and which paid for doctors and staff. She also asked me if I would volunteer on Tuesday and Thursday nights. At first I was reluctant—homelessness wasn't one of my issues—but how could I say "no" to Alina? Once I started, I actually enjoyed it. There was a core group of medical students who came every night—clearly our most dedicated—and lots of patients with interesting problems. After Joe had some difficulties finding a full-time physician, I suggested I'd be willing to work there if he'd also hire Phil. Phil and I could work as a team and bring in medical students to work with us during their primary care clerkships and as evening volunteers. I had received a large grant to move primary care education into the community, and Camillus Health Concern seemed a good place to start. Moving my office into the clinic was a radical move for a tenured professor. I told Margaret I'd have to give up my sessions in special immunology. She and I had been the only ones of the original group left. In truth, I was no longer needed. Her research grants allowed her to fund several new faculty positions, all specializing in AIDS.

We had 10,000 charts in the clinic. Each chart represented a patient who had been seen at least once, a life that had fallen through the cracks of society. Some, like Jennifer, were mentally ill or severely addicted to alcohol or crack cocaine. Others were merely poor or were the victims of bad luck.

The patients coming to the clinic reflected Miami's diversity—black, white, multiethnic, Latino—with one notable exception: There were very few Haitians. In fact, in the six years I worked at the clinic, I had only two Haitian homeless patients. One was a lawyer with bipolar disorder who was too proud to tell his family that he had lost his job. Instead, he just disappeared into the streets. The other had AIDS but didn't want his family to know. I asked the social workers at the clinic how they might explain this. After all, Little Haiti was by far Miami's poorest community. Their consensus: First, there was a very low incidence of alcoholism and drug use among Haitians. It wasn't just that they couldn't afford the alcohol or drugs; their culture didn't condone them. Second, family was the ultimate Haitian safety net. If you fell on hard times, someone in your extended family would always take you in. Finally, as a last resort, returning to Haiti was preferable to life on the streets.

The clinic offered comprehensive services to homeless people, including HIV counseling and testing. Every month Ruth, the head nurse, would give me a copy of the monthly report on counseling and testing activities. For several months they passed in front of my eyes but did not register in my conscience. Then I started to notice.

"Uh, Ruth, are these numbers right?" I asked her as she passed by my open office door.

"Which numbers, Art?"

"The counseling and testing numbers for January, February, and March."

"What's the matter with them?"

Ruth was a meticulous person who carried the day-to-day activities of the clinic on her shoulders. She took pride in the counseling and testing program. No other clinic for the homeless had one, and she had developed it from scratch. She stopped making copies at the Xerox machine and came into my office. If there was a problem, we needed to fix it immediately.

"The numbers testing positive seem awfully high. Seven out of 40 for January, 10 out of 53 in February, 12 out of 42 in March. That's going to average somewhere between 15 and 20 percent."

"That's what we've been averaging since we started testing. Is this the first time you've read my report?"

"I must have made a mental error with my decimal points. But could this really be true? Nobody has numbers this high. Fifteen to 20 per thousand is high, but 15 to 20 per 100? Could there be a bias? Do only those who feel sick ask to be tested?"

"I don't think so, since the Centers for Disease Control gets pretty much the same numbers from their seroprevalence study. The interesting thing is that it's pretty much all heterosexual transmission. I'll show you." She got up, went to her office, and returned with more data. "You see, we have very few gays, and surprisingly few IV drug users. The prevalence in men and women is essentially the same."

"What's this 'sexual assault' category?"

"Oh, those are guys who had nonconsensual sex in prison. But even if you factor that in, it's still mostly heterosexual sex."

"How do you explain it?"

"Crack cocaine, sex with prostitutes, lots of unprotected sex. The price of a 'basic service' by a street prostitute is the same as a 'nickel bag' of cocaine—$5. That's not an accident. For the drug dealers, crack is literally money that grows on trees. They give a huge discount for volume. Sex and crack, the opiates of the homeless. I saw oral sex performed for bus fare—75¢—right on the sidewalk in front of the clinic the other night. I surprised them as I was leaving. Sex is one of the few things with a value that they can't take away from you when you're homeless. Lots of our guys work—in the labor pool, crushing cans, windshield washing. It's not enough to rent an apartment, but it can buy you a few minutes of release. Fifty pounds of aluminum cans will get you five bucks and that will get you laid. Unfortunately, the cheaper the prostitute, the more likely she has the Ninja."

I thought about what Ruth was saying and intuitively knew she was right. I knew many of the prostitutes she was talking about, since they frequently used the services of the clinic. They were not Hollywood-image prostitutes. No Julia Roberts look-alikes. They had been burned out, frequently abused, and physically changed by the pathological effects of their addictions. A few were men in drag. Some gave up prostitution when they found out they were infected with the AIDS virus. Others continued selling themselves to support their habit or merely to keep from starving. Other women were not strictly speaking prostitutes but bartered sex for crack cocaine or simply got screwed while high. Crack became so popular in Miami because it was cheap and gave a powerful high, and the word on the street was that IV drug use could give you the Ninja. Crack is smoked, not shot up. Once addicted to crack, few cared about the future anymore, so every day and night the games of Russian roulette were repeated in alleys, vacant buildings, and on old mattresses all around the clinic. We had over 100 patients known to have AIDS that we followed on a regular basis. Fortunately, there were enough minor victories among these patients to keep us going.

Juan got a job as a house painter. He was one of a special group of patients. He discovered he had AIDS while recovering from drug addiction. There were 20 such patients living at the shelter at any given time, as part of a substance abuse recovery program. They got a single room, worked in the kitchen or on "crew," supported each other, and eventually got set up with a job and an apartment of their own. I was proud of Juan because he didn't cave in and start using again when he found out he had AIDS. Four other members of his group were in a similar situation. He wanted to use the time he had left to make peace with himself and feel good about himself. He didn't look ill. Many recovering addicts turn to weight lifting and other sports to build their self-esteem and help them cope with withdrawal. Juan was barely five feet two inches tall, but had a physique resembling Arnold Schwarzenegger's. His platelet count was low, and he had to take medicine to keep from bleeding or bruising spontaneously. It took five pages of forms and four visits to various social agencies to get him to qualify for zidovudine, under the newly funded Ryan White program. Politically astute legislators had named their ambitious program to provide treatment for everyone with AIDS after a well-publicized boy with AIDS and hemophilia—an "innocent" victim. Hemophiliacs need lots of concentrated blood products to stop them from bleeding, and before a blood test for HIV was developed, many received blood products that had been contaminated by the virus. I would later discover there was a bizarre Haiti connection to poor Ryan White. The Miami company that manufactured the concentrated product (called cryoprecipitate) needed lots of whole blood. Where better to harvest it but in Haiti, where there were plenty of people desperate enough to give until they passed out, at $10 per unit? Some of that blood was contaminated with HIV virus. The contamination of cryoprecipitate with HIV wiped out a generation of hemophiliacs.

A year of zidovudine cost about $3,000. Usually by April the program that funded zidovudine for poor people in Miami ran out of money, and we couldn't enroll new patients until July. Juan found a job. He moved into his own apartment. He "adopted" one of our older patients with cancer, took him to the hospital for his radiation treatments, and translated for him there.

Far and away, our biggest success was Jackie. If ever there was a person who played to all the stereotypes of the AIDS epidemic, it was Jackie—a six-foot-two, 200-pound black gay transsexual. "Her breasts are bigger than mine," said Ruth, somewhat jealously.

"It's the estrogen she takes" was my comeback. "If you had that much estrogen flowing through your veins, your breasts would be huge, too."

Jackie was well known to the clinic even before I started working there. She was one of the first patients Ruth had persuaded to be tested for HIV. Not only did she test positive, but her blood counts said she was not far away from full-blown AIDS. She had been living on the streets for years, prostituting herself to support her drug habit—not just the usual crack habit but extraordinary amounts of estrogen, the feminizing hormone. The ultimate outcast and proud of it, she would come to the clinic wearing a blonde wig, a tank top, no bra, the shortest of skirts, and high heels. She was all about attitude. Once I slipped and referred to her as a "he." Her stare was withering. But she needed us. Without estrogen she could not be the woman she wanted to be. I was her only legal source for estrogen. Ironically, in Miami it was getting harder to get estrogen on the street than it was to get crack or heroin.

"Jackie's here to see you, and only you, Art," announced Ruth. "I think she's in love with you."

"Now Ruth—you've just got issues with her cup size."

"Well, it just doesn't seem fair."

In addition to HIV, Jackie had high blood pressure. She adamantly did not want to take AIDS medicines and insisted on taking one and only one medicine for her blood pressure. She knew she was different—I'm sure her attitude was a defense mechanism—but we had leverage: She really wanted estrogen.

"Look, I'm not going to prescribe estrogen to someone who's living on the streets, prostituting herself, maybe infecting others with HIV just because you've got body image issues," I stated, matter of factly, in the semiprivacy of an examining cubicle. "On the other hand, I'll meet you halfway. Kick your habit, come in for regular visits, and give up your life on the streets, and we'll work with you."

Jackie was used to bargaining with clients but never about her health. "You would do that for me? I could come here rather than going to the Jackson Clinic? I hate it there. People are always staring." As she said this, she herself was staring down at her own cleavage.

"Sure! Let's get the ball rolling. Start by talking to one of the social workers. We'll get you in a drug program and follow your HIV infections here. You're in luck. Someone actually donated some Premarin to our pharmacy. Who knew? I never thought we'd use it, but why not use it for you?"

I had taken a basic negotiating skill we teach our students—establishing a therapeutic contract—and applied it to a homeless, black, HIV-positive, prostituting transsexual. It worked. She came every month. I knew that in the beginning she came for the estrogen alone, but slowly, surely, she started to take her health seriously. Curiously, her blood counts did not deteriorate appreciably from month to month. It was almost as if she had partial immunity to the effects of the virus. "You'd better watch your blood pressure," I told her during one visit in which she confessed she had been lax about taking her medicine. "Mark my words. Something other than the AIDS virus is going to do you in if you don't take care of yourself."

For every temporary gain or spiritual healing, however, there seemed to be 10 medical and spiritual failures. For example, I had to find a way to tell Charisse she would die soon. She and her two children had been living on the streets. Her family threw them out when they found out she had AIDS. She felt well, but her T cell count was down to 30. At that level, death is usually just a short time away. "Have you made plans for your children?" I asked when the medical student assigned to her faltered. One of them was with her, a waif under 2, playing obliviously with her shoes. He most likely had AIDS also. "You're at a stage of your illness where you will become ill soon. It's time to think about these things." Charisse burst into tears. The student looked as if she might cry also. "No, I haven't made any plans. I'm just going to Hell. I guess HRS [Health and Rehabilitative Services, Florida's welfare program] will take my children."

All Good Things

AND THEN THERE WAS SAM. As Sam loosened his overcoat, his hidden machete, held to one side by the cinch of his coat, slid down his pant leg. The tip buried itself into the rubber tile floor of the examining room with a musical ching. The fact that Sam was wearing an overcoat in the heat of Miami's summer was the first clue that he had once again lost touch with reality. His chief complaint—"I want to kill somebody"—was the second. A gentle man when clean of cocaine, a homicidal paranoid schizophrenic when smoking crack, he had relapsed once again. Weapons, of course, were not allowed in the clinic. Carrying a concealed weapon into the clinic would result in Sam's being banned from services for two years. My immediate concern was to make sure no one got hurt.

Fortunately, there was little real privacy in the examining cubicles, separated from the corridor only by a curtain that was impossible to completely close. I positioned myself on a stool where Ruth could see me. The machete ching caught her attention, and I was sure she could see it balanced against Sam's shin under the curtain. I hand-signaled "911" to her while I engaged Sam in mindless banter. That was her clue to call the police; clear the students, staff, and other patients out of the clinic; and station security on standby in case Sam suddenly turned violent. My job was to keep him calm until the police arrived.

"What's up, Sam? Long time no see." I ignored the machete.

"The FBI put a radio in my head. I can't turn it off." Sam was distracted by a bug crawling under the sink counter.

"Well, I can see that would be a turnoff," I deadpanned.

"Huh?" said Sam, half listening to me and half listening to the voices in his head.

The police arrived in a matter of minutes. They hid around the corner of the examining room until they heard Sam launch a particularly long and disjointed sentence. Before he could get to the direct object, they were on him, one officer hammer-locking his head and arms, another grabbing the machete.

"Nice work, Doc," claimed the arresting officer.

"Not bad for an academic," I replied.

"Where's my Baker Act?" In Florida the Baker Act allows a physician to treat a mentally ill person considered to be a danger to themselves or others involuntarily.

"Gee, I've been kind of busy. I'll have it for you in two minutes."

Ruth flashed me her "if anything had happened to you I would have killed you" scowl. Not waiting for the verbal zinger sure to follow, I retorted: "Hey, poor Sam's slowed by drugs, mental illness, and AIDS. If I couldn't outsmart him, I would have outrun him!"

"It's a good thing he didn't come in while ABC News was here! Sam's too scary. Our 15 minutes of fame would have been reduced to 10 seconds or less!"

ABC had been by a few weeks before at the insistence of Joe, my former resident who founded the clinic. They put together a nice piece on the problems of the homeless, with some great sound bytes from the dean on how we were developing a "culture of compassion," a small clip from my interview, and a lot of footage of Joe in his white coat and our medical students working out under the bridges. Conspicuously absent was Phil, who, in reality, was doing most of the work. He managed to find himself needed as far as possible from the camera on the day of the interviews.

"What did you think of the news clip?" I asked him offhandedly the morning after it aired on the evening news.

"It will be the death of our work here," was Phil's surprising answer.

"What do you mean?" I responded, stunned by his candor.

"That video went right to Joe's head. Just watch. We're no longer going to be about health care for the homeless. We're going to be about Saint Joe."

In fact, Joe's presence at the clinic was a rare event in the weeks prior to the news story. He was completing his fellowship in hepatology and planning on entering private practice. After the story we saw more of him for a while but always with an entourage of reporters or wealthy patrons.

As a result of the news Joe generated about the clinic and our medical students' heroic work after Hurricane Andrew, the University of Miami School of Medicine received the first-ever Community Service Award from the American Association of Medical Colleges. By the time the award was received, however, we were already sliding toward the dissolution of all we had accomplished. A new clinic administrator told us of changes mandated by the federal government. We'd have to reduce losses by enrolling patients in Medicaid and billing our patients for their copayments. Phil was beside himself with anger. Always outspoken, he told everyone who'd listen of the ridiculousness of billing homeless people. It was soon clear that Joe and the new administrator did not want to hear this, and Phil was targeted to be pushed out. In a preemptive strike, I reassigned Phil to a new free clinic we were starting in Overtown, a historically black community a few blocks away. Without Phil, my work at Camillus became impossible. I left a few months later.

A week before I left, Jackie came to see me. "I heard you're leaving, Dr. Fournier. Who's going to take care of me?" she asked as she gestured dramatically, with her hand pressing against her breast bone.

"There'll still be other doctors here," I responded.

"But they don't know me like you know me. Inside out, if you know what I mean. What will happen if I don't get my estrogen? I'll just shrivel up and lose all my curves! Listen, I've got an apartment now and insurance—both Medicaid and Ryan White. Can I be your private patient?"

I still had one half day each week during which I saw private patients. I liked Jackie's idea. She would be the one patient I would take with me from the homeless clinic. Managing her problems really was too complicated to turn over to someone else. Besides, we had made a contract and she had kept her end of the bargain. She had kicked her drug habit, was no longer prostituting herself, and was no longer living on the streets.

"You know Jackie that just might work. Here, take this number and call for an appointment. I'll need to see you again to repeat your blood tests in four weeks."

Jackie kept her appointment. In fact, I followed her in my private practice for 12 years. She came faithfully every month, so that I could follow her HIV infection and renew her estrogen prescription. She was one of the first patients to become a long-term survivor of HIV infection—all told 15 years without requiring medicines. The office visits became like social calls. She would flirt in an innocent way and socialize with my office staff. When she died in 2004 (of a health issue that was not AIDS), I was both saddened and relieved. I had been faithful to the end to this most challenging patient.

I viewed leaving Camillus Health Concern with ambivalence. On the one hand, the clinic really had raised the consciousness of Miami concerning its homelessness problems. In a moment of enlightenment, the citizens of Miami-Dade County voted to tax themselves to establish a comprehensive approach to homelessness. As a result, the number of homeless people in Miami has been drastically reduced. And we succeeded in giving our students a window on the human condition that few doctors in training ever experience.

On the other hand, with so many social problems—drugs, poverty, mental health issues—it was hard to convince myself that the health care services we were delivering amounted to much more than a Band-Aid. I asked myself over and over: *Are we making a difference? Could we ever make a difference?*

PART II

Secrets Revealed

Titid

October 15, 1994 I HAD FINISHED SATURDAY ROUNDS and was looking forward to going sailing. While changing clothes in my bedroom, I turned on the television to catch the weather forecast. To my surprise, a special event was being broadcast. President Aristide was returning to Haiti. The television station was flipping back and forth between Little Haiti, in Miami, and Port-au-Prince. I sat down on my bed, arrested in the process of changing clothes by the spectacle unfolding before me. The plane carrying President Aristide back to Haiti had just landed in Port-au-Prince. He stepped off the airplane, waving to dignitaries and American troops. When the camera flashed to the presidential palace, tens of thousands of Haitians were outside, awaiting the president's return. Then the camera switched to Miami. Haitian-Americans were dancing in the street. Joy was everywhere.

Back in Haiti, President Aristide entered one of five helicopters that would simultaneously depart for the palace. For security reasons, no one was told which helicopter would land first. I watched as the five helicopters left within a minute of each other for the short trip between the airport and the palace. The crowd outside the palace roared when the helicopters appeared. President Aristide left his helicopter and entered the palace. Several minutes passed, and the cameras focused on the crowds pressing outside the palace fence. The military estimated the crowd at 200,000 people. The masses of people in the center of the crowd pressed the people on the fringe against the iron bars of the palace fence. They didn't seem to care. They were singing and waving palm fronds. When President Aristide finally stepped in front of the bulletproof glass podium, the crowd exploded with noise and song.

The camera flashed back to Miami. The Haitian-Americans continued to dance in the street, chanting "Titid! Titid!" The president's words were soft and difficult to understand, as he first spoke in French, and then in English. He could say anything and it wouldn't matter. Democracy had returned to Haiti. I was transfixed by the event. A few hours passed, and I was still sitting on the bed, watching the television. I didn't go sailing that day. The Haitian embargo was over, and I had witnessed it on the television screen.

I had become somewhat acquainted with Haitian politics over the 14 years I had been caring for Haitian-American patients. That had been a particularly turbulent period in Haiti's history. The decadence of "Baby Doc" Duvalier and his exploitation of the Haitian people ended in 1986 when he fled the country under threat of a popular uprising. He was succeeded by a series of equally undemocratic and repressive military men until 1990 when, to everyone's surprise, a parish priest from Cité Soleil, the poorest slum in Haiti, won the first fair election in Haiti's modern history. "Titid," an affectionate Creole abbreviation that means "little Aristide," campaigned on the premise that the poor were the overwhelming majority and that it was about time they ran the country. "Peace in the heart, peace in the belly" was his motto. He survived three assassination attempts during the time of the generals. Nine months after his election, the army staged a coup and sent him packing to Venezuela.

American Democrats, particularly key members of the Black Caucus, loved Aristide. Republicans believed his liberation theology was uncomfortably close to communism. Fortunately for Titid, Bill Clinton was president. In response to the coup, a two-year economic embargo was orchestrated by the United Nations. When that failed to dislodge the junta, an armed intervention led by the United States finally restored Aristide to power. No wonder his return created such a spectacle. For two years the Haitian peasants had endured shortages, price gouging, falling tourism, and the black market, all to get their Titid back—their leader, perhaps even their savior. And now, thanks to some U.S. Army helicopters, there he was, back in their midst.

At the time, my younger daughter, Suzanne, was in the ninth grade at a private school in Miami. I shared carpool responsibilities with my neighbor, Barth, the newly appointed chair of neurosurgery. The following Monday I was passing time with Suzanne in the driveway, waiting for Barth and his two sons. When Barth arrived, he rolled down his window.

"Do you have any interest in Haiti?" he asked.

"Yes!" I responded immediately.

"Come to a meeting in my office on Wednesday morning. We are going to get into Haiti in a big way."

"Do you mind if I bring some of our Haitian-American faculty?"

"No, bring whoever you want. I've got some doctors from Haiti who are coming up who want to work with us."

"I'll be there," I answered, as I threw Suzanne's book bag into the trunk.

As they drove off, it struck me as odd that Barth would have any interest in Haiti. He had recently started the Miami Project to Cure Paralysis. *Surely that must be a full-time job in itself,* I thought, not to mention the challenges involved in running one of the medical school's most prestigious departments. I later learned that Barth's good friend from college worked as a missionary in Haiti. His friend's brother was rumored to be the fourth-wealthiest person in Texas. Barth had asked the family for a sizable donation to his paralysis project. The family complied but with a condition: Barth had to get involved in health care in Haiti.

I arrived at Barth's office 10 minutes early the following Wednesday. Three men had arrived before me. One was American. The other two were speaking French. All three were subdued, with somber expressions on their faces, as if they were in church. "You must all be here for the same meeting that I am," I said in French. Danny T. introduced himself in English as the director of a nondenominational Christian missionary group that works in Haiti. Barth's friend from college was his boss. Marlon and Jerry were twins, distinguishable only by the fact that one had a Shakespearean beard, and the other a goatee. I spoke to them in French, learned from a combination of college classes and a vague recollection of lullabies from my French Canadian grandmother. Upon hearing my French, their expressions changed to broad, relaxed smiles.

"These fine young Haitian doctors want to invite you to come to their country to see what the conditions are there, so that your school can help them," Danny T. explained. It seemed as if he had known them for years. I later learned he had met them for the first time just before I arrived.

Barth was over an hour late, but it mattered little. We were soon joined by Henri, one of our Haitian-American faculty members, and Junia, a Haitian-American student who had worked for me while applying to medical school. Junia was from the same part of Haiti as Marlon and Jerry. With the arrival of these two, the conversation became even more animated and changed into Creole. Marlon and Jerry speak English but were thrilled to have the opportunity to tell their story without the struggle of having to translate it and were delighted to find other Haitians at the meeting. There were times they were speaking so quickly and simultaneously that I couldn't follow what they were saying. Other times, by focusing on the French words I recognized, I could get their meaning. Marlon had trained in obstetrics and gynecology and Jerry in internal medicine. Both, however, spent most of their time performing surgery, since there was a critical shortage of surgeons. The embargo and politics under military rule had left their hospital in ruins. They wanted our help in rebuilding it and an opportunity to come to Miami to learn better operating techniques.

When Barth arrived, the discussion switched back to English. Prior to the embargo, Barth had come up with a plan to provide medical assistance to Haiti with his college friend, but the plan had been put on hold during the embargo. "The need is even greater now," the twins told us. Agreement was quickly reached. There was much that our school could do, but the first step would be to put together a team to visit Haiti and assess exactly what the problems were and what we could do about them. The trip would be in December, just two months away.

That's how "Project Medishare" was born. At this time, doing international humanitarian work was not a traditional role for an American medical school, even one with as strong a tradition of community service as ours. And candidly, our school, as is true of so many medical schools, was struggling to meet the demands for service in Miami. It was hardly in a position to take on the daunting challenges of Haiti. We would need an independent charity to raise funds and support volunteers until we had enough of a track record to compete for grants and major donations. Barth and his college friend chipped in the start-up costs and asked me to put together a team to go down to Haiti and figure out how we could help.

As the meeting officially ended, the twins, Henri, and Junia informally lingered with much laughter, smiles, hugs, and cheek kisses. It seemed like a family reunion, even though they had just met. The twins' goodbye to me was, "Dr. Fournier, we'll see you again in Haiti!" At dinner that evening I announced to my wife and daughter Suzanne that I was going to Haiti in December.

"No, you are not!" my wife scolded.

"Yes, I am," I calmly replied.

This interchange was repeated at least 10 times. Suzanne looked back and forth silently between the two of us as if she were watching a tennis match.

"Barth put you up to this, didn't he? He is going to suffer. Wait until I tell Kathy [Barth's wife]. What are you going to do there?"

"Meet the doctors there, see what the conditions are like."

"Who's going to pay for it?"

"We'll raise the money from contributions."

"Who'll do the morning carpool?"

"That can be your contribution."

"Why would you want to go there, anyway? They've got diseases on top of diseases. It's not safe. They'll put a tire around your neck and ignite it."

"That won't happen. Don't be silly. You can't stop me. You might as well accept it."

"Just make sure your insurance is paid."

So first there was AIDS, and now there would be Haiti. My professional and personal lives seemed to be spiraling in ways I could not control. I blamed myself for my part in the article that stigmatized Haitians for having AIDS. Beyond that I was haunted by memories of Régis and all my other Haitian patients and shamed by the fact that I hadn't been able to do more for them. Janet would probably never understand. After 14 years of involvement, I had to go. Somewhere down there, perhaps I'd find some answers. Perhaps I could make up for past mistakes.

Port-au-Prince

December 1994   THE FLIGHT FROM MIAMI to Port-au-Prince is only an hour and 20 minutes. As we taxied down the runway, preparing for takeoff, I began reading Internet reports about Haiti and its problems. Haiti, once the richest colony in the French empire, is now the poorest nation in the Western Hemisphere. Founded after the world's only successful slave rebellion, it only recently became democratic. The reports made for grim reading: average income, $175 per year; life expectancy, 57 years; 75 percent illiteracy rate; and extraordinarily high death rates from diarrhea, tuberculosis, and malaria—all treatable diseases. I lifted my head from a report, looked out the window, and saw that we were already over Haiti. It was a day so clear that even from 30,000 feet, I could make out trees and huts on the land below. Haiti is a mountainous country, and the deforestation described in one of the reports is painfully apparent as you approach Port-au-Prince from the air. The capital city sits in a valley between two mountain ranges next to the mouth of a river. The gray of the city contrasts with the green and brown of the mountains and the blue of the Caribbean Sea. Despite the deforestation, my first impression was that of a rugged but beautiful land.

At Barth's request I had recruited a team to visit Haiti to see the situation firsthand and assess what we might do. The team consisted of 23 doctors and nurses, most of them faculty members at the University of Miami. In addition to Barth and myself, the doctors in the group included Lynn, chair of family medicine; Michel and Henri, two Haitian-American faculty members also from family medicine; Ron, the Haitian-American director of our radiology residency program; and Jackie, University of Miami alumna who directed Miami's public health department. On the nursing side, Diane, dean of the University's school of nursing, and Lydia, one of her faculty members, were interested in international nursing from an academic perspective. Their scholarly approach to the problems faced by Haiti's nurses was balanced by the practicality of Ruth, the head nurse of Miami's homeless clinic. Junia, my assistant, and some photo and video journalists Barth had recruited from the *Miami Herald* and our local Channel 10 rounded out the group.

During the flight down, Danny T. visited each row of members of our group and gave instructions:

"Stay together. Pay no attention to requests to take your bags, even if someone looks like they are in some kind of uniform. And above all, don't submit to the temptation of giving something to people begging, especially the children. It will start in the airport parking lot, and if you give to one, you will have to give to everyone. Then you will have nothing left. Be prepared for an assault on your senses. Haiti is a land of contrasts."

As we flared over the runway, rows of U.S. Army helicopters came into view, along with groups of U.S. soldiers and their equipment. "It looks like we are flying into a war zone," someone murmured.

We exited the plane and descended the portable ramp to the tarmac. We were immediately hit by a blast of heat and wind. The trade winds spilled over the mountains to the east and rolled across a broad plain toward the Bay of Port-au-Prince. Impressive mountain ranges rimmed the plain to the north and the south. A small *konpa* band played as we entered the terminal.

Following Danny's advice, we got through customs and baggage claim swiftly. Danny T. was right. The begging started in the airport parking lot, as we loaded into the *Pajeros* that would take us to our hotel. Each of the four small trucks was surrounded by 10 to 20 children, mostly boys, ages ranging in age from 8 to 14. In English we heard, "You got money for me?" "You help me, mister?" "Quarters?" "Dollar?" These pleas were made softly and frequently with a smile. Our driver responded harshly in Creole but without deterrent effect. We finally began to drive off, with children clutching our open windows, half of them running and half being dragged along with us.

The assault on our senses had begun. The air was filled with the acrid stench of burning garbage. As we proceeded from the airport to the center of Port-au-Prince, we could see piles of uncollected trash smoldering by the side of the road. Some of these heaps were taller than I was. Some covered a half acre. Their blue smoke, mingled with the dust blowing off the naked hillsides, stung our eyes and obscured the clear blue sky we had seen from the airplane. Even in the capital, goats were nibbling on trash, live chickens hung by their feet in open-air markets, and burros were used as beasts of burden. For that matter, so were humans. Small men pulling two-wheeled carts piled high with boxes were everywhere. Women carried water, food, or bread on their heads. We passed a military checkpoint on the way downtown. With that one exception, we saw no sign of law, nor any sign of disorder. The narrow streets were crowded with people. In fact, it seemed as if the entire city lived in the streets. Automobile traffic was heavy and slowed by the crush of people. However, a gentle toot of the horn and pedestrians immediately yielded the right-of-way to the automobiles. There were no traffic lights working, but our drivers were skilled with their horns as they approached intersections or corners and were careful to negotiate the seemingly endless series of potholes, puddles, and ruts. If we stopped our minicaravan, children appeared from the crowd and asked for money. They were gently discouraged by our drivers. For the most part, though, we were ignored. *How strange for such a poor country to have so much industry in its streets*, I mused. People were weaving hats and baskets from palm fronds, making rattan furniture, and refurbishing old box springs, all out on the streets. Brilliantly colored paintings hung from walls and fences.

"A nation of artists," Henri volunteered, "but who's going to buy their works?"

Most of the traffic consisted of large trucks carrying construction materials or "tap-taps," Haiti's form of public transportation. Tap-taps (named for the sound made by their engines running) ranged from small pickups outfitted with a carved wooden "bonnet," through buses, to 12-wheelers. For a few gourdes one could travel all around the city on the smaller pickups or, for a few gourdes more, all the way to Cap Haitien or Jérémie. The larger tap-taps have their route painted on the side, front, and back. Both small and large are decorated with bright colors, paintings, designs, and frequently hand-carved wood trim. They also have elegantly painted religious phrases or *pawòl granmoun,* expressions of folk wisdom. Most spewed out diesel smoke that mingled with the smoke from the piles of burning garbage we encountered every few blocks, to sting our eyes and make us pinch our nostrils. It seems there had been no trash pickup during the three years of the embargo, and people burned their garbage with gasoline or kerosene when the piles became insurmountable obstacles.

The color of the streets varied from drab concrete gray to explosions of bright colors, particularly around the open-air markets, with a seeming abundance of squash and vegetables accented by the white feathers and red combs of the upside-down chickens. Everyone (in most places the sidewalks were five to six people deep) seemed to have some place to go or something to do. Toothless old people mingled with groups of school children in uniforms, walking as a group with knapsacks on their backs.

We stopped at the Holiday Inn before going to the university hospital. In the back of the hotel was a courtyard, a tropical garden with shade, a breeze, and leaves that seemed to soak up the haze, dust, and diesel smoke. Those of us who were in Haiti for the first time shared our initial impressions. For our Haitian-American faculty, it was the beginning of a four-day reunion. Some had not been home for more than a decade, and all had not been able to return home during the three years of the embargo. Every new face in the hotel came over and introduced himself or herself, followed by smiles and embraces.

Marlon and Jerry rejoined us. They would be inseparable from us over the next four days. They were in their element now. Any shyness that had afflicted them in Miami had melted away in their own country. They were ecstatic that we had come and hopeful that we could help them and their people, even if that prospect for help was in the distant future. They were classic twin brothers. One started a sentence and the other finished it. Their English was excellent when they were fresh, but when they were tired they sometimes struggled for words. They were kind enough to use English, since most of our group did not speak French or Creole. They carried small radio phones, which were constantly ringing, and they had the uncanny ability to carry on one conversation concerning patient care in Creole with each other while speaking to us in English. Their energy was boundless and their dedication obvious. Several times over the next four days, one or both would excuse themselves "for just a brief emergency surgery." Then they would disappear for 20 minutes to half an hour and then reappear and resume the tour. At the same time, they had the kind of surgical irreverence that made "M*A*S*H*" so humorous. They wanted to take us on a tour of the university hospital before the officially planned tour the following morning. As we passed by a nursing class, several students waved, called their names, and blew kisses.

The university hospital had been built by the U.S. Marines during their occupation in the first half of the 20th century. It reminded me of the hospital in which I had done my residency two decades ago—a complex of two- and three-story buildings with barrel-tile roofs and open-air porches. Across the street from the entrance to the hospital was a collection of drugstores, mortuaries, and coffin makers.

As we were entering the emergency area, we passed two older Toyota vans with red crosses painted on their sides. One had its hood raised, and the other had a flat tire. "Our ambulance fleet," commented Marlon dryly. Between the two defunct ambulances and the small door labeled "Emergency" was a tap-tap with two legs dangling outside. The woman inside had had a seizure and had yet to regain consciousness. She was being tended to by her fellow passengers. We entered a dark corridor. The walls were painted a dingy olive drab, and the stale smell of urine permeated the air.

"This is the ER stabilization area." The room we were in had several small cubicles. In one a resident was quietly suturing a laceration. In another cubicle a two-day-old baby with a colectomy was nestled in a crib with his mother at his side. Jerry explained that the baby was born with an imperforate anus. He and his brother had performed a colostomy to save the baby's life.

We proceeded through a maze of dimly lit corridors and passageways connecting one ward to the next. Each ward contained 20 to 30 beds. They were divided according to services—medicine, orthopedics, surgery, pediatrics, obstetrics. Each ward had more patients than there were beds. On most wards this meant that patients were doubled-up, two per bed. In the orthopedics ward, though, because of traction and other hardware requirements, there was only one patient per bed. The mattresses on the beds were old and stained, and there were rarely sheets. There were no nurses, no toilets, no running water, and no medicines. Marlon explained that if they wanted a patient to have medicine, they had to write the patient a prescription, and either the patient or a family member would have to take it to a pharmacy across the street and return with the medicine to the hospital. At that point the patients were responsible for taking it themselves. Food and nursing care were provided by families. Those patients without families depended on passersby or fellow patients.

Flies were everywhere in the surgical ward, where burn victims were recuperating. Burns are common in Haiti, where most people cook over open fires. Children frequently burn themselves when they pull over pots of food suspended over the fires by tripods. The burn victims avoided the flies by sleeping under mosquito netting. There were probably six or seven burn victims there that day, their dark uninjured skin in sharp contrast to the pink, yellow, and mottled burned tissue. Flies buzzed around a container filled with antiseptic solution. Jerry described how, without a sterilizer, unsterile gauze pads were dipped into the antiseptic solution to pack the wounds of surgical patients. Eighty percent of the emergency surgical patients developed sepsis, and 60 percent died. Many patients seemed to be there because they had no place else to go. Their doctors could figure out they had terrible maladies but lacked medicines to treat them. One woman with a tumor growing out of her skull had been waiting five months for surgery. James, a young boy with a beautiful smile, skated down the corridor on tin cans strapped to his knees. His legs were permanently folded at the knees by scarring that had developed after severe burns. He had been waiting two years for surgery, living at the hospital the entire time. There were some patients who were young but gaunt. *These patients must have AIDS,* I thought to myself. No one mentioned them or their problems, however. In fact, our Haitian hosts seemed to ignore them.

Our next stop was an orphanage in a place called Post Cazeau, a complex of newer buildings on a large plot of land, near the airport. We could see planes taking off and landing just beyond the treetops. Marlon and Jerry said they would like to build a new hospital in a vacant field next to the orphanage some day. Miriam, the director, volunteered that she would donate the land. She also had land out in the countryside, in a village called Pestel, that she would donate to Project Medishare.

Miriam trained as a nurse and has been doing missionary work in Haiti for 23 years. She and Marlon and Jerry formed a "mutual admiration society." In addition to running the orphanage, her organization, World Harvest, sponsors open-air clinics in villages in the countryside.

"As bad as things are in Port-au-Prince, they're 10 times worse in the countryside," she informed us. "When we hold one of our clinics, 200 to 300 people will come from all around. Children with malnutrition and typhoid fever, people with tuberculosis and malaria. In Pestel there are 60,000 people in the surrounding area and not one doctor. The government has a building there, but there's no one to staff it. We're so pleased you're going to help."

Marlon and Jerry pulled off a minor miracle by inviting the minister of health, the president of the International Red Cross in Haiti, the director of the university hospital, the dean of the medical school, and several other important people in Haitian health care to meet us at Post Cazeau. We later learned that our visit was the first occasion that brought all these dignitaries together in one place and time. In retrospect, however, I must admit this diplomatic miracle did not lead to long-term results.

Miriam, her staff, and the orphans set up tables in a field next to the orphanage, under a canopy of palm fronds to provide needed shade. Our Haitian hosts were in agreement with regard to their priorities. First, the university hospital needed to be resuscitated as a teaching hospital. Then it needed to train doctors and nurse practitioners who could provide general medical care. Finally, there needed to be a coordinated system for transportation around the country and a way to maintain medical equipment and supplies. Project Medishare, representing, at least in the Haitians' eyes, the University of Miami School of Medicine, could help in all areas—by donating medical equipment and supplies, by developing the training that was needed, and, of course, by raising money.

As a group of teenagers played homemade instruments and sang
Haitian folk songs under a tree the entire afternoon, Miriam provided lunch and a tour of the orphanage. The orphans sang songs
and presented Barth with a bouquet of flowers in honor of our visit.
Miriam told us the story of each orphan in the chorus. Most were
survivors of the wreck of the Neptune, a ferry boat from Jérémie that
capsized during a storm. While the incident received fleeting coverage in the American press, it had become legendary in Haiti. Every
street artist had a painting of the wreck of the Neptune, full of visual
imagery of drowning innocents.

The children, some 60 all told, were well behaved and affectionate. They made friends with each and every member of the group.
Miriam told us that each one was hoping for adoption. Most of the
faculty completed the tour with a child holding on to each hand.

I had been warned by some Haitian friends in Miami to beware
of the missionaries. Many missionaries are seen by the Haitian people
as agents of the Central Intelligence Agency, and some missionaries
have come to Haiti with pejorative attitudes about Voodoo or the
Haitian work ethic. Much of the ceremony at the orphanage would
have struck a cynic or skeptic as a calculated play for sympathy, but I
felt skepticism draining from me in the face of such obvious need. In
the course of our meetings, Miriam agreed to take six babies who
had been abandoned at the university hospital. Without good nursing care, these children rarely survived long in the hospital. Any
thoughts on my part that Miriam might be grandstanding were
washed away by the heartfelt gratitude of the Haitian hospital officials, for whom the abandoned babies represented a huge problem.

We reentered our *Pajeros* and snaked slowly in silence up the hill
to Petionville, where our hotel was located. Although it was late at
night, the life in the streets still pulsed. Although Petionville is considered a well-to-do suburb, there were people begging on the street

outside our hotel. Parked immediately next to the hotel was an abandoned hulk of a burned-out car, with five children living in it. Several members of our team kept bringing food from our meals out to the children. The hotel itself, built in Caribbean gingerbread style during a brief surge in tourism in the 1970s, was an oasis of comfort and service. This only served to further subdue the group. Some were too emotionally moved after the experiences of the day to enjoy the hotel's hospitality. Others went directly to the bar.

One doesn't have to spend a long time in Haiti to realize that social class is a huge issue. I had read about this unique part of Haiti's history as I prepared for our trip. The French plantation owners of Saint Dominique were true aristocrats—counts and marquises—but only French men were willing to give up the luxuries of France and endure the hardships and risks of the tropics. So they took slaves as concubines. The progeny of these liaisons were treated by their fathers as a special class. They weren't free, but their fathers acknowledged their paternity and allowed them to own land, receive an education, and manage the plantations. So the Haitian revolution was really two revolutions rolled into one—a slave uprising by Africans, taking advantage of the chaos created by the French revolution to win their freedom, and an Oedipal revolt by the lighter-skinned *mulattes*. Both classes joined forces to drive out the French. Shortly thereafter, however, conflicting visions of who owned the country erupted into civil war. The result was a stalemate, with the Africans and their descendants controlling the land in the countryside and the *mullates* controlling the institutions—government, church, and education. The *mulattes* became, in effect, an unlanded aristocracy. Class politics became an endless cycle of power plays, oppression, and revenge.

These class tensions have dominated the history of Haiti since the revolution—one class believing that Haiti is their birthright, the other that only they have the knowledge and skills to manage the country. Each is simultaneously dependent on and suspicious of the other. These tensions were readily apparent at the bar at the Kinam

Hotel. The staff, coal black, talked among themselves in Creole but responded to their patrons in French. The bar clientele were all light-skinned, fashionably dressed, and insistent on impeccable service. Their eloquent Parisian French shamed my French-Canadian patois.

I pulled Michel, one of my Haitian-American colleagues, aside to a quiet corner and asked him, "What's the attitude of the elite toward the poor here?"

Michel himself is a very light-skinned Haitian but more of a member of the intellectual elite rather than the economic elite. He traced part of his ancestry to French Huguenots from Nantes, and in truth the only hint of Black Africa in his features is the black curliness of his hair. He showed me on a map of Haiti where the plantation bearing his ancestral name was located. However, Michel was the rare individual with the ability to step outside class and take a truly objective view. He thought for a while before he responded. "I'd have to say there's a spectrum of attitudes that range from 'noblesse oblige' to 'let them eat cake.' You can live well in Haiti, even if you're not particularly rich, if you come from the right family. The poor provide servants—cooks, maids, and chauffeurs. Some are treated as members of the family; others are frankly abused. I've always found it ironic that a country born of the only successful slave revolt in the world has so many of its people working as servants. For them it's almost like the revolution never happened. For the most part, however, most of the elite are ignorant of the plight of the poor. They're isolated from the problem. They never leave the capital."

"The gulf between the rich and poor is pretty glaring," I remarked. "The houses on the crest of the hill above the hotel are as impressive as anything in Miami and with a much better view. Isn't it a little ironic that we're down here trying to help and some of the elite could care less?" I asked.

"It's not that simple, Art," Michel responded. "Some do care. Charity is different here, however. A family might quietly support a poor family they know or one particular orphanage. Occasionally, a prominent family will adopt a cause like a hospital or an orphanage.

Most of the doctors are sons or daughters of the elite, and all of us give back to a greater or lesser extent. Look at Marlon and Jerry. The thing is, you're seeing all this with the eyes of an American. I wonder really if it's all that different in the United States, except there's more wealth and less poverty there. But take everything we've talked about with regard to 'class' and substitute 'race,' and that's what you've got in the States. How many white Americans feel any responsibility or sense any connectedness to black Americans?"

Cité Soleil

THE FOLLOWING MORNING WE returned to the university hospital for the official visit. I spoke with our Haitian-American faculty as we assembled outside prior to the tour. All three were stunned by what they had seen the previous day. "Haiti always was a poor country, but the health care system worked," said Michel. "I trained here in medicine. I had a good experience. Now, there is nothing left. It's like a shell."

"I went by the radiology department yesterday," added Ron, who ran our radiology residency program, "and none of the machines are working. I can't tell you how much things have deteriorated."

"What happened?" I asked.

"Number one is the embargo. Remember, while humanitarian supplies were allowed in, things like mattresses and sheets didn't count. Also, under the embargo, the price of everything went up, which promoted theft. Things could always be sold on the black market. Second, politics influences everything in Haiti. Under the military government, all appointments were political. People could just go to the ministry of finance and pick up their paycheck and then go to their real job. There was no accountability. Finally, in the days just prior to the U.S. intervention, there was the revenge factor.

The military was not about to leave anything of value to the Aristide government, so they took what they could and trashed the rest."

We entered the building and were led to the office of the newly appointed director of the hospital, where we also met with several members of the faculty and medical staff. The director was quite candid. "We have nothing!"

After introductions in English and French, and a brief statement from each of us as to what we hoped to accomplish, we embarked on the tour. To Marlon and Jerry's surprise, with the exception of the presence of one or two nursing student volunteers on some of the wards, the official tour showed us pretty much the same conditions as we had seen the day before.

After the tour, our caravan weaved through the pedestrian traffic from the center of Port-au-Prince to St. Catherine's Hospital in Cité Soleil.

"Cité Soleil is the lowest point in the city," Dale, our driver explained. "It was where they unloaded the slaves in colonial days. During the Papa Doc years, they changed the name to Cité Simone, in honor of his wife. They resettled people from the countryside during those years. It was never intended to be permanent. Now, there are people who have lived their entire lives there."

The neighborhoods we passed through were clearly poor, with small, one-room homes built of concrete blocks. Many were unfinished, doorless, windowless echoes of life in the poorest country in the hemisphere. Once, crossing over a river of open sewage, I asked, "Is this Cité Soleil?"

"Not yet," responded Dale.

Ten minutes later, we took a left turn. The narrow road dropped 10 feet in elevation, flattened out, and plunged straight ahead into a seemingly endless plain of cardboard and tin shacks. Footpaths snaked between the shacks, and wisps of smoke rose from the open fires used for cooking. The smell of urine and feces intermingled with smoke and the aroma of cinnamon and vanilla. There was even more life in the streets—buying, selling, sidewalk industry—than in

Port-au-Prince. People wearing rags and naked children bathing in washbasins or chasing each other between the shanties mixed with people in starched white shirts, lace-trimmed dresses, or school uniforms. "This is Cité Soleil," Dale informed us.

The only other motorized vehicles on the road were trucks and Caterpillars hauling out piles of garbage. The government had made Cité Soleil its first priority for garbage removal. "How silly," I thought, since the whole encampment was built on a pile of mud, rocks, puddles, and trash. Whenever the process of trash removal slowed our progress, children again surrounded us, gently asking for money. The adults occasionally smiled and waved but, in general, paid no attention to us, as if it would be impolite to stare at such strange people in such strange vehicles. Although human voices sounded everywhere, without the noise of the automobiles, there seemed a strange aura of peace in Cité Soleil.

It was even quieter in St. Catherine's Hospital, which rests behind a wall in an enclosure that includes a school and an orphanage. It's located in the further reaches of Cité Soleil, a short distance from Port-au-Prince harbor. From the second story of the maternity ward, we could see children playing and women washing their clothes in a sewer lake the size of a football field. In fact, one can see all of Cité Soleil from that vantage point, a tribute to its flatness, the low height of its huts, and the compactness of the area, home to 200,000 people.

St. Catherine's was run by a private group under contract with the Catholic Church. Compared to the university hospital, it was functional. Here, there were nurses in the wards, clean sheets on the beds, and a food service. Most touching was the maternity ward. Most births in Haiti take place at home, but conditions in Cité Soleil were so unhealthy that St. Catherine's promoted "intensive" prenatal care and in-hospital delivery. There was still crowding, with two mothers in each bed, 60 mothers all told, along with their newborns. In St. Catherine's the newborn child was kept with its mother, who nursed it as needed. I was struck with the contrast to our nurseries in the United States, with their plastic bassinets, confining blankets, and bottles of formula. *Why should a newborn spend the first days of its life mostly in isolation?* I thought. Three sets of twins were born the day we were there. One set was nursing simultaneously, one on each of their mother's breasts. Perhaps there were things we would learn from Haiti, despite its problems, rather than assuming that coming from the United States we knew it all.

"This is better. Don't you think?" I said to Ruth, the nurse from Miami's homeless clinic, as we were leaving. Touring St. Catherine's, I didn't notice that she had dropped out of the tour in the maternity ward or that she had been crying.

"Better?" She whispered incredulously. "What hope for life could these children possibly have? You saw what we drove through to get here. They'll stay in the hospital for two days and then go home to what? A dirt floor, a cardboard roof, and an open fire? It's so sad. They have nothing."

I invited Ruth to join the team because I hoped her experience running the homeless clinic in Miami would translate into practical advice as to how to address primary care and nursing issues in Haiti. Based on our prior work at Camillus Health Concern, I knew I could count on her for no-nonsense answers.

"They have their mothers' love, and their fathers', too." My words convinced myself, but I'm not sure they convinced Ruth.

"Get real, Art," she came back at me. "This is one notch above hopeless."

That afternoon, on the way back to the hotel, our caravan ventured down one of the few paved side streets in Cité Soleil. The *Pajeros* stopped so that the journalists from the *Herald* and Channel 10 could take pictures of children playing in the open drainage. There was not a tree or a blade of grass in Cité Soleil, only people, animals, shacks, and trash. Our vans were surrounded by 30 to 40 children, laughing, smiling, begging, and reaching through the open window to touch us.

I spoke my first words in Creole: *"Nou pa genyen lajan"* ("We don't have any money").

"W'ap ban mwen valiz-la?" ("Will you give me your purse?")

These words, almost sung in a sweet soft voice, came with a smile from a young girl with an angelic face. She pointed to Ruth's pocketbook wedged between us.

"She'd like your pocketbook, or its contents," I translated. I was just starting to crack the code between French and Creole.

"W'ap ban mwen valiz-la?" the girl repeated. She was almost nonchalant, with her head nesting on one arm, while the other reached forward to stroke Ruth's leg. She was pressed against the van by the other children behind her but did not seem to care.

"Ou trè bèl" ("You are very pretty").

"Mèsi, msye" ("Thank you, sir").

"Kijan ou rele?" ("What's your name?").

"Regine, msye" ("Regine, sir").

I wished for a moment I could take her and all her friends home with me. Some had the red hair and swollen bellies from malnutrition, but most were apparently healthy and some stunningly beautiful.

"W'ap ban mwen valiz-la?"

When she realized we wouldn't give her the purse, she became content to stroke Ruth's dress and leg with her extended finger. She appeared about 12, her budding breasts partially exposed by her tattered T-shirt. Her companions on my side of the car asked for the four Dixie cups we had used for drinking water and that were kept in the back of the van. When I gave them the cups, the lucky four ran off to play with them in the drainage, followed by an envious entourage. Regine and several others stayed with us. I wondered how many other strangers she had begged from. How unreal to be sitting in the middle of Cité Soleil, surrounded by 200,000 of the poorest people I had ever seen. I had more money in my wallet than any of these children's parents (if they had parents) earned in a year—maybe 10 years. And yet there was no sense of danger. In fact, there was a mystifying aura of peace, fostered by the absence of cars and the softness of the children's voices. Then a dark thought entered my mind. Anyone who wanted to could take advantage of these children. Any one of them, boy or girl, could probably be had for as little as a quarter.

I thought back to our article about AIDS among Haitians and the controversy it had generated. While our critics were wrong in claiming that our patients were all gay, perhaps there was a grain of truth in the assertion that gay men had been coming to Haiti and giving money to young boys for sexual favors. It probably was not just a gay thing, though. There were probably plenty of straight men making the short flights from Miami or New York to buy sex from Haitian girls and women. Exploitation has no sexual orientation. There had been much debate in the medical journals and the press as to whether the virus evolved in the United States and was transmitted here or vice versa, or in some entirely different place (we now know it probably originated in Africa). But seeing Cité Soleil, I could only wonder what the point of the debate was. Wasn't it just another way of fixing blame? There had already been too much blame, I suddenly realized, and not enough understanding. If anything, it was politics, economics, and exploitation that spread the virus. For that we're all responsible and we're all to blame. The majority just loves to pin the blame on a minority, particularly one that can't fight back.

As I sat in the van, surrounded by squalor, another thought struck me. The temperature was 90 degrees. The sun was heating up the water seeping through the garbage, and every possible combination of genetic material—human, viral, and bacterial—was fermenting in the water percolating through Cité Soleil. Children and adults were washing in, drinking, and excreting this water, walking or playing in this soup, sometimes with cuts or open sores on their feet.

My God, we've created the world's largest Petri dish, I thought. In this soup of human and non-human DNA, anything could evolve. I never for a moment actually believed my fanciful Petri dish theory but was intrigued with it nonetheless. How much easier it would be to fight AIDS if only its origins were unrelated to sex. AIDS could have started here, in Lagos, or perhaps in Bangkok. It didn't matter where it started. Poverty was the issue, not sex, and we could do something about that.

The Petri dish theory did have an ominous corollary. The more the virus replicated, the more opportunities it had to evolve new strategies to kill people. If, at the present time, one couldn't get AIDS from splashing in or drinking contaminated water, there was no guarantee that it wouldn't happen in the future. In their quest for survival, germs have proven remarkably ingenious. For example, the germ that caused the Black Death evolved two ways to infect humans: one through the bites of rat fleas and the other from person to person through the air we breathe. The rats and their fleas, coupled with the masses of people that crowded European ports in the 14th century and the crowded living conditions they endured, set up a double-barreled killing machine that wiped out half of Europe.

But the real Petri dishes were the tens of thousands of people living here and in places like here—walking culture media—who don't know they're infected and couldn't do anything about it anyway. So, whatever its origin, a virus has evolved that's smarter than we are, smarter because it attacks our three greatest weaknesses—our immunological defenses, our sex drive, and our social order. And as Tim had predicted, it always stays two steps ahead of us. Back in the States we're still debating about confidentiality, voluntary testing, and condoms. How can you expect safe sex to be practiced in Cité Soleil or, for that matter, any place that's desperately poor, or by the mentally ill, or by crack addicts on the streets of Miami? The real risk factor is not being gay or Haitian or a drug user. It's being a social outcast. AIDS is like a zombie curse, a judgment cast on a victim. People with AIDS are victims of a disease, and we blame them for it. Then they exist, half-alive, half-dead, enslaved by their diagnosis. And we, as a society, are not going to do anything about it because we must want it this way. Otherwise, we would change things. And why do we want it this way? At best it's because we're all wrapped up in our own lives, and who has time to care about places out of sight, like Cité Soleil? At worst, our inertia must spring from an attitude that "someone's got to win and someone's got to lose." The Haitians seemed to be losing at every turn. To my way of thinking, even if you didn't give a fig about these poor Haitians and were only motivated by enlightened self-interest, the Petri dish corollary said it would be a big mistake to tolerate conditions like those in Cité Soleil and not try to change them.

All these thoughts raced through my mind in the short time it took for Regine to run her finger across Ruth's leg. I thought of my own daughters. What accident of birth or stroke of fate destined them to a life of comfort in Miami and Regine to be begging and perhaps prostituting herself to strangers in Cité Soleil?

"Wake up," I told myself. "You can do something about this!"

"I wonder what happened to our journalists," murmured Ruth, looking over her shoulder out the back of the *Pajeros*. The crush of children and Regine's stroking her leg were unsettling to Ruth. Each passing second made it more and more difficult to resist Regine's persistent pleas.

The journalists finally reappeared. "Unbelievable shots!" they exclaimed as they piled into the van. That one trail carved out by that trickle of water led us further and further in, and it got darker and narrower, with no end to the people. Seven or eight in each shack—that sort of stuff."

"I'll bet AIDS is a problem here," I said to no one in particular.

"They say in some parts of Cité Soleil that everyone has AIDS," one of the journalists responded.

It was almost impossible to make a U-turn in the narrow street we were in. In the process our driver knocked over a tripod pot containing a soup or stew. That brought out from the owner the only words of anger I heard during my first trip to Haiti. After changing direction, we slowly drove away, with the children running after us, half singing, half shouting, *bonswa, bonswa!*

That night, back at the hotel, I allowed myself the luxury of a bottle of wine after dinner. I sat on the terrace sharing the day's events with the Haitian-Americans in our group—Michel, Henri, Ron, and Junia. I discovered we all shared the blessings of a Catholic education: a broad base of philosophy, theology, and history, with the church's particular spin. My education and Junia's, guided by the Augustinian fathers in the United States, had emphasized philosophy and theology. Michel, Henri, and Ron, taught by the Jesuit fathers in Haiti, were strong in history, language, and politics. Michel could translate both Latin and classical Greek. He laughed at how hard he had studied those subjects for so many years. "What relevance do Latin and Greek have to life in Haiti?" He asked. "But the good fathers believed we needed a classical education." Junia mentioned how hard it was going to school in Miami. Her parents had immigrated so that their children could get an education. Her schoolmates teased her relentlessly about her language and her clothes. All lamented how many traditional values were lost in the process of Americanization.

I confessed my role in the Haitian AIDS study. They all looked at me with an expression that said, "Art, how could you have done such a thing?" Unfortunately, the study was infamous for the political fallout it had caused Haitians. I realized that all my colleagues had experienced the unspoken accusation—"You're Haitian. You must have AIDS."—and I was partly responsible for that.

After only two days in Haiti I had come to realize the absurdity of lumping all Haitians together. Here was Michel, light-skinned progeny of professional-class parents, sitting beside Junia, coal dark, whose father died when she was young and whose mother worked as a custodian. And then there was Régis, my "blood brother" dentist, and Regine, the young girl in Cité Soleil. What did they have in common? Yes, they were all Haitian, but that was about it. So why would being Haitian put you at risk for an infectious disease? How naive we had been.

I explained how my thinking had changed, about my thoughts in Cité Soleil, the Petri dish theory, the zombie curse theory, and, most importantly, what Ruth and I had learned from Regine. All remained silent for a while, pondering the gravity of my assessment. We then launched into a serious discussion of all the factors that might drive poor people anywhere, not just in Haiti, to sex early and often, the status of women and their dominance by men, prostitution as a survival strategy, family breakup for economic or political reasons, emigration, lack of education, lack of health care, and lack of hope for the future.

Then one of the journalists interrupted to tell us that a famous Haitian singer was there in the hotel and was going to sing for us. Beethovan Obas and his brother Mozart gathered us together in another part of the terrace. His stage name when spoken, not as spelled, in Creole means "Down with Beethoven." His music was a blend of jazz and island beat, sometimes in French, usually Creole. The sound was soft, upbeat, and happy. But if you understood the words *"Ayiti Mouri"* you would see that they mean "Haiti is dying." As he sang, the Haitian faculty joined in the chorus: *"Nou pa, nou pa, nou pa, nou pa . . ."* (We're not, we're not, we're not, we're not.") He was left-handed and played his small acoustic guitar upside down, inventing new cords and finger formations. We sang until after midnight, until an American businessman came out of his room and screamed that we were keeping him awake. Junia, Henri, Ron, Marlon, and Jerry decided they would take Barth, Ruth, and the others out to a Haitian disco. "How can you go dancing after what we saw today?" I asked them.

"Haiti is like an onion, Art," philosophized Henri, solemnly. "You've got to peel away one layer at a time. What we saw today is why we dance."

Lost

I AWOKE THAT MORNING AT 4:00. My mind was full of images and ideas, visions of the people in Cité Soleil and the hospitals and plans of how we could help. Not wishing to disturb my roommate, Lynn, I silently slipped outside the room and went to the pool to recline on a chaise lounge. The people who had been sitting around the terrace a few hours earlier had completely disappeared, some to their rooms and some to dance through the night. The night staff at the hotel had curled up on couches and were fast asleep. There was no electrical power in Petionville that night or, for that matter, in much of the rest of Haiti. Since the United Nations-sponsored embargo to restore Aristide, the capital averaged four hours of electricity each day. As a result, the stars were brilliant and plentiful in the clear skies above. It was a surprisingly cool evening, with a gentle breeze and no mosquitoes.

The tranquility ended with the sound of voices singing in the street. *Why would people be singing at 4:00 in the morning?* I asked myself. To see over the wall that protected the hotel terrace from the street, I had to climb the stairs to the second-story balcony. Coming up the hillside street next to the hotel was a group of five women, all balancing 20-gallon buckets of water on their heads. One woman would sing a line, and the other four would respond in chorus. It was a happy, joyous song, like gospel music, except in Creole rather than English. Earlier in our trip I had noticed the head wraps—the *twòkèt*—all the women in Haiti wore. They were suggestive of turbans, and the group coming up the hill demonstrated to me their purpose: It made it easier to balance heavy loads on their heads. I thought back to Ginette and her seminar on Voodoo and the zombie curse. Now it was becoming clear, as I listened to the women singing. Here the French god and the older African gods lived side by side, but the old gods were closer to the people and maybe even more helpful to them.

"Haiti is a land of contrasts," Danny T. had warned us on the airplane. How strange to be in a place only an hour and a half from my home that seems like it's on a different planet, a place where not even water can be taken for granted. I had noticed women carrying water into Cité Soleil. When washing their children, they would have the children stand in a basin and pour the water over their heads so they could collect the water and use it again. Now as I watched the women climbing the hill, it was clear that even in Pétionville water was a very valuable commodity. Yet in a country where no one had collected the garbage for the past three years, everyone was washing their clothes by hand and then hanging them out to dry on shrubs or tree branches. People worked and played, suffered, and enjoyed life simultaneously.

At 4:30 the roosters began to crow. Oddly, this reminded me of home, since I lived within earshot of Little Haiti and always arose early. At 5:00, the church bells began ringing and the scent of freshly brewed Haitian coffee permeated the air. It was Sunday. I had lost track of time. The new god was calling. In Haiti people go to church not once but twice on Sunday—once in the morning and once in the evening. Of course, the second ceremony could be Voodoo.

At 6:00, I decided to leave the hotel to watch the sunrise. Even at that hour there already was an old, toothless woman standing just outside the door, begging. To the right, about a block away, a footpath zigzagged up a steep hillside. There was much foot traffic going both up and down. I wondered where the path might lead, and guessed it might provide me with the best sunrise opportunity.

The path was no more than a yard wide. If you passed someone, you both had to turn sideways. The people coming down the hill were a mixture of families going to church and people carrying water or vegetables to the Pétionville open-air market. Many of these people were barefoot. I wished I was barefoot, since the steepness of the path and the slipperiness of my leather soles forced me to frequently use my hands to assist in my ascent. People coming down the path would offer me their hands in assistance. It was hard for many to conceal their bemusement at a *blan* climbing the path, and yet everyone smiled as we passed and everyone said "*Bonjou.*"

"*Bonjou,*" I responded, and returned their smiles.

The top of the path revealed a dirt road that encircled an old stone wall, resembling a fort or castle. The night before, Michel had told me how, after they had expelled the French in 1804, the Haitian leaders built forts and castles in high strategic positions throughout the country, anticipating that the French would return to try to recapture their prize. Haiti, the land of contrasts. The only successful slave revolt in the history of the world and after 200 years it had yet to escape the slavery of dictatorship or foreign control.

The view from the base of the fort was spectacular. I could see the entire city of Port-au-Prince and the entire valley the city sits in, turning into countryside as it extended to the east. To the north and to the south were long sweeps of coastline with rugged mountains. In the city three things dominated the view—the presidential palace, the cathedral, and the cemetery. Above the wall the mountains rose perhaps another 2,000 feet, dotted with the homes of some of Haiti's wealthiest families.

The elite had taken a lot of criticism in the American press for its role in supporting the dictatorship. One writer had even coined the phrase "morally repugnant elite," or MRE for short. The attitudes I heard expressed in the hotel bar toward President Aristide—sometimes stated openly, sometimes veiled in political correctness—shocked me. In the American press during the embargo, Aristide was portrayed as a folk hero—a saintly, scholarly man with the courage to stand up to the dictators. Now, in the postintervention politics of both Haiti and the United States, the press at times and the opposition continuously presented a different picture. Aristide was just another demagogue, with his own circle of henchmen and assassins, all out to get rich in the process of controlling the government, a new aristocracy of the formerly poor. What had I gotten myself into? How much of this was true, and how much represented class-based politics? As an American it was best not to pass judgment, at least until I knew more.

Focus on our humanitarian mission. Stick to health care. Don't take sides. Stay out of politics, I reminded myself as I thought back on my experiences of the day before.

Stop trying to lay blame on anyone. Just accept the fact that Haiti is yin and yang—a culture with two sides—one elite, book-educated, light-skinned, Catholic, French, and urban; the other poor, tradition-educated, dark-skinned, Voodoo, African, and rural. On my descent down the path back to the hotel, I extended a helping hand to those ascending, as had been done for me a short while before.

Back in Miami, Brother Paul, who ran the homeless shelter, had suggested to Ruth that while in Haiti she should try to find Father Luc and Brother Rene, two Brothers of the Good Shepherd who were working in an orphanage in Port-au-Prince. Our only knowledge of their location was a scrap of paper on which Brother Paul had written, "somewhere just off Demas 31, Frères Bon Berger" and a phone number. I volunteered to go along. Titi, our Haitian driver, said he knew Père Luc and that he was a good friend. He did not, however, know exactly where he lived. On Sundays more life spills into the streets of Port-au-Prince than even during the workweek. People are going to or coming from church or are simply out for a Sunday walk. Everyone is dressed in their best clothes. The churches were overflowing with people, spilling down the stairs leading to the

A market woman and her *twòkèt*. Photo by Wassim Serhan.

entryway. We found Delmas 31 easily and headed down beyond Grace Children's Hospital, looking for a sign that said Bercail Bon Berger. Titi, however, was not looking for signs. He looked to the left and to the right, searching the crowd to find Père Luc. He would occasionally stop and call over to a passerby. *"Èske ou konen ki kote kay Pè Luk?"* ("Do you know where Father Luc's house is?"), he would ask. Everyone to whom he directed this question at first looked puzzled. Then they would turn 360 degrees. Some shrugged their shoulders. Others claimed to know Pastor Luc and seemed excited at the mention of his name. No, they didn't know where he lived, but it was just up ahead and to the right or to the left. We continued this process for several miles, hoping we were getting closer. Finally, we met a young man who said he knew where Pastor Luc lived. After a brief conversation in Creole, he climbed into the van with us and gave directions that led us off of Delmas 31 into a quiet residential area.

"*Sa se li*" ("This is it"), said our guide as we disembarked. We knocked on the door and were greeted by a young mother and four children. Unfortunately, it was not the home of Père Luc and Brother Rene. We asked her if she had a telephone. She didn't, but she walked us four houses down the street to the home of one of her neighbors who did. She went inside, spoke with her neighbor for a minute, and then they both came out to invite us all in. I had Titi call the number on the scrap of paper we had from Brother Paul while we talked with the families. He spoke rapidly with someone on the other end, and I couldn't understand everything he said. After he hung up, he assured us that he had just talked with the housekeeper of Père Luc and knew exactly where she was. We thanked both families and piled into the van to head back in the general direction of Delmas 31. We took encouragement when we saw "Bon Berger" painted on a garage door. We stopped and inquired from some people on the street, who identified the house across the way as the home of Luc's housekeeper. One of the people in the group decided to accompany us, so now there were five of us. We met the former housekeeper for Père Luc and Frère Rene. Unfortunately, the priests had found their building too small for their needs, so they had moved to the other side of Delmas. She volunteered her husband to guide us to that location. The neighbor from across the way would guide us to where her husband was.

The five of us piled into the van and headed back toward Delmas. We pulled over next to a small restaurant/bar and waited while Titi pounded on the door. After a few minutes, a slight man came out, exchanged greetings with the entire entourage, and traded places in the van with his neighbor. He gave directions to Titi in Creole as we headed back up Delmas and crossed over to its other side. He was sure the new location was at Delmas 42. Alas, there was no sign of Père Luc. Someone in the crowd there spoke English with a New York accent, having grown up in Brooklyn. He said he thought Père Luc might have taken his orphans to church. So in one last desperate attempt, we doubled back to one of the churches looking for either Père Luc or his orphans. The two gentlemen who had now been accompanying us for several hours climbed out of our van in front of the church and made their way through the crowd inside. A few minutes later they came out to tell us that indeed they had seen the orphans but that Père Luc was not there. I turned to Ruth and suggested that it was time we call off the search. We had been looking for five hours. I suggested to Titi that we needed to bring our guides back to their neighborhoods. He said no, it wasn't a problem. And no sooner had he said that than they ran across the street to catch a tap-tap back to Delmas 31. We had taken up several hours of their lives, but we barely had time to thank them. As we headed back to the hotel, I noticed that tears were once again streaming down the corners of Ruth's eyes.

"I'm sorry we didn't find Father Luc and Brother Rene," I offered in consolation. "Next time we'll need to get a more accurate address."

"It's not that," she quivered. "It's just that people have so little here and yet they are so helpful. Five hours of their lives to help us find someone. Can you imagine that happening in Miami? And then they wouldn't even let us take them back."

Back at the hotel we shared our adventures with some others in the group. "That's because politeness is taught in the schools here," Ron informed us. "I was taught, and indeed everyone is taught, if you ever come across a stranger who has lost their way, you stay with them until they find where they are looking for."

That evening at the hotel, even Lynn, the most senior member of our group, got caught up in the dancing. He and some others decided to take a walk before dinner. Near the market square in Pétionville they were caught up in a *ra-ra*—a surge of people singing and dancing, part political, part religious, part festive—that swept up the streets from Delmas. The costumed musicians and dancers of the *ra-ra* surrounded the group and demonstrated their dance steps. Lynn responded with a meringue he had learned in Haiti 40 years earlier. The journalists captured the festivities on video. I stayed behind at the hotel again, talking to the bartenders, Junia, and Henri.

"Why is President Aristide so popular?" I asked in Creole after a few drinks. Junia and Henri were quiet, so quiet that I knew I was guilty of another breach of diplomatic protocol. But perhaps because I asked in Creole, one of the bartenders volunteered an answer. "Here's what you need to understand. Papa Doc was feared but obeyed. Baby Doc was hated because he was such a *gwo manje* ('big eater'). Titid stood up to him and the generals. Anyone who survives three assassination attempts, God is with him."

Secrets of the Zombie Curse

Our last day in Haiti OUR HOSTS WANTED TO TAKE us out of the city. Life in rural Haiti is very different from life in the capital. Our destination was Kenscoff, about 15 kilometers away, on the other side of the mountains that rim Port-au-Prince. We ascended quickly from Pétionville, through a partially canopied road. Trucks heading in the opposite direction were filled with workers returning from the mountain quarries. In valleys and ravines we could see clusters of huts and the smoke of open-air fires. On the road there was a thin but steady stream of people—women heading to town on burros, or with baskets on their heads, or men herding goats or carrying sheaves of wood. We crested one mountain ridge, descended, then started up again in a new valley we could not see from Port-au-Prince. The countryside was dotted with small wood-frame homes decorated with wood carvings and palm-thatched roofs. Farmers tilled their fields, and in the distance long lines of women could be seen descending footpaths, carrying produce on their heads. Kenscoff was about three-quarters of the way up the valley, with a deep ravine and terraced slopes on the opposite side running all the way up the mile-high mountain. Just before Kenscoff, there was a Baptist mission, a museum, a clinic, and an orphanage. Across the road was a collection of leather goods and paintings for sale by local artisans.

The museum contained more than I expected. In addition to Arawak pottery and slave-carved furniture, there was a collection of toys artfully constructed from trash by the children of the mission, which is the oldest in Haiti. We met the foundress, who, with her husband, had started the mission and clinic 49 years earlier.

"You see the terraces on the hillside? The mission helped the people build them, so they could farm," explained Danny T.

"Unfortunately, they didn't hold during tropical storm Gordon last summer," the foundress added. "Seven hundred people were washed down the ravine. Some were never found. The saddest thing, though, were those who climbed out, covered with mud, who lost everything. We still have seven families we haven't found homes for. They just live outdoors—foraging for food. If you find someone in the U.S. who wants to do something, we'd appreciate your help."

There was irony in her comments about the flood. There had been a small item in the Miami newspaper about 700 people drowning in Port-au-Prince. Either these people in Kenscoff were totally ignored, or the reporter didn't know the difference between the city and the countryside. As a second irony, that day the news reported that 1,500 people died in an earthquake in Japan. That tragedy made front-page news for a week.

We toured the grounds, including the clinic, where they used a butcher's scale with an attached sling to weigh the children. The clinic was busy, with a line of people outside the door waiting to be seen. Orphans, identifiable by their shared uniforms, were running around chasing each other. We had lunch in the American-style cafeteria. The menu was mostly hamburgers and hot dogs. I had no interest in eating, but I did order a plate of french fries and sat down with Ruth at a window with a view of the whole valley.

As we were preparing to leave, Henri and Junia gave me books as a gift for making their trip possible. One was a book on learning Creole, and the other was a book of Creole proverbs. The foundress gave me a toy helicopter made of oil containers and bottle caps and a book entitled *God Is No Stranger*. The text contained spiritual expressions from the people of Haiti. The photos were beautiful—candid, but capturing the souls of their subjects. To my surprise, on the twenty-fourth page was the same woman I had seen in a photo in Régis's book 13 years before—the one sitting in the chair having her tooth extracted.

This was your mission, wasn't it Régis? It had to be. It was the first, and the only one here long enough for you to grow up in and to return as a dentist. Is this where you returned to die? Are you here in one of the tombs I saw as we ascended the mountain? Or did you feel too stigmatized to return home? Did you die in the hospital in Port-au-Prince? Are you buried in that huge cemetery I could see from the hillside in Pétionville? Is it a comfort that I still think of you? It has been 12 years and not a week goes by that I don't think of you. You and all the others. But especially you, my blood brother. You were infected by the blood of one of your patients. The first occupational fatality. The first to die as a consequence of your professional duties. Am I the only one who knows the truth? You should have been hailed as a hero. Instead, they insisted you were gay, so they could blame you for your fate, wash their hands of your blood. If you were an innocent victim, there might be other innocents as well, and God would have no justice or mercy. But you were all innocent. We have no one to blame but ourselves—our own ignorance and willingness to tolerate intolerable conditions. You were just the most irrefutable case, and it took me 13 years to understand. It's bad enough that you all had to die, but how much did we add to your suffering through stereotype and blame? You had a curse on you, didn't you? But, then, so did I. Cursed by naivete and enslaved by conventional thinking. I've been sleepwalking through the biggest medical event in my lifetime, enslaved by the constraints of my world and ignorant of the reality of yours. It took this visit to your home to wake me up. Once you see it, it all makes sense—the poverty and the things some people have to do to survive in places like Cité Soleil. So even if we can't cure AIDS, we ought to be able to do something about its cause. The problem is, most of my

colleagues have never ever seen anything like this. They're under a zom-
bie curse that's even worse than mine. No point in worrying about that
now, though. I've got to focus on what I can do, how I can make a
difference, Régis. I promise I'll try.

Ruth caught me staring across the valley, lost in thought. She touched my arm. "Time to go, now," she spoke softly. "What were you thinking about?"

"Just daydreaming. I knew a person from here once. A dentist. He died of AIDS. One of the first." I was still staring across the valley.

"How do you know he was from here?"

"He told me so, in a roundabout way. Are you glad you came to Haiti?"

"Glad? It's been the most moving experience of my life."

"It's good to hear that. I was a little worried. Every time I turned around, you had tears in your eyes."

"It's just that I've learned so much, in such a short period of time, and it's been so emotional. I have to confess, I came here with some preconceived notions, and they've been pretty much shattered."

"You're exceptional in that regard. Most people have their minds permanently made up. Most people in the United States have never seen anything like Cité Soleil. Even if they did, it wouldn't change the way they think for more than a minute."

"That's a pretty dark view of humanity for such an optimistic person, Art."

"It gets worse. You remember that young girl in Cité Soleil who wanted your purse?"

"Remember her! I'll never forget her."

"What if I told you she would sell herself for money. Maybe it hasn't happened yet. Maybe it won't happen for a few more years. Maybe it won't happen to her, but it could happen, if not to her, then to some child just like her."

"Having seen Cité Soleil, I wouldn't blame her."

"You see, you are the exception. We blame the poor for their fate. It's the only way we can justify in our own minds the fact that we live so much better than they do. The same is true for all our outcasts and pariahs. Have I ever told you the secret of the zombie curse?"

"The zombie curse? Where did that come from?"

"It has to do with AIDS, both here and back home in Miami. But not just AIDS. Lots of other stuff, too. I'll tell you about it as we ride to the airport. We really do have to go."

Marlon and Jerry were at the airport to see us off. They were already planning our next visit. "You'll come back, won't you, Dr. Fournier? Come for Carnival."

"*Se sèten, Marlon ak Jerry. Ayiti te met yon wanga sou mwen*" ("For sure, Marlon and Jerry. Haiti cast a spell on me"). I put this phrase together from the book Junia had given me a few hours before. It took Marlon, Jerry, and all the people in our group by surprise, and they broke out in laughter. I needed a laugh myself. If I couldn't dance, at least I could laugh. Haiti had cast a spell on me. But it had also freed me from the zombie curse that had enslaved my mind all these years. A magical land. A magical people. A land of contrasts. An onion, and I'd only peeled away the first layer.

"I must come back."

Medishare, Inc.

THE INVERSE CULTURE SHOCK of returning home to Miami started even before we landed. Out my window I could see row after row of neat suburban houses, most with swimming pools. Sunlight reflected off the windows of the skyscrapers downtown, while cars streaked along the expressways. I felt like I was returning from another world.

Janet and Suzanne met me at the airport. As I entered the car, Janet started talking about home issues and things that had transpired in our family while I was gone. I could tell by her body language and facial expressions that she didn't want to bring up the trip, fearing (correctly) that it might have changed our lives forever.

So it was Suzanne who, 10 minutes into the trip home, first inquired, "Well, how was it, Dad?"

"Words fail me, Suz. Imagine a place where you can't even count on clean water or electricity, where the children can't count on going to school. Yet the people were beautiful, and there's so much we can do to help. I'll tell you one thing, though. We're never going to take anything we have here for granted ever again."

Something about my comment about not taking things for granted set Janet off. She launched into a diatribe that filled the rest of the trip home. How dare I imply that she and Suzanne didn't appreciate everything we had in Miami, how hard she and I had worked to get what we've got, how frugally she managed our budget and how modest our means were. She wasn't really talking to me but to Suzanne. What if, corrupted by her father's radical notions of altruism and social justice, Suzanne decided she wanted to go to Haiti herself?

Janet's fears were not totally off base. After dinner, while I was watching the news in our den, Suzanne came in, cuddled next to me, and asked me to tell her about the trip. I told her about the hospital, the orphanages, and life in the city and in the countryside. Mostly I told her about the people, particularly the children—the orphans and Regine and the others in Cité Soleil. She listened silently. When I was done, she kissed me on the cheek and said, "When I'm a little older and I actually have something I can contribute, I'd like to go to Haiti with you."

Janet's attitude softened over the next few days, the thaw fueled in part by Suzanne passing on my stories to her in bits and pieces and also by the positive press we received. Channel 10 did a nice three-minute clip on our trip, and the *Herald* did several articles on various aspects of health care. We were famous for three days; long enough for friends and family to call, congratulate me, and gently admonish Janet not to be so hard on me. Still, after each of these conversations, she would always ask, "You're not going back there, are you?"

Shortly thereafter, Barth, Michel, and I set up a meeting with Bernie, dean of the University of Miami School of Medicine. While we were in Haiti, we debriefed every evening, brainstorming as to how we could help. Our students and residents could come as volunteers to orphanages almost immediately. Just doing screening exams and creating medical records would be a huge benefit for those children. We could do health fairs like the ones we do in Key West, only modified for Haitian health issues. We had met many Haitian doctors—Marlon and Jerry, for example—who wanted to come to Miami for further training. Michel was particularly interested in introducing training in family medicine, as this discipline did not exist in Haiti. But to accomplish any of this, we would need the dean's blessing.

Bernie had been the dean for my entire tenure on the faculty. He had been instrumental in helping me get my first big grant and supported my academic advancement. He had coined the phrase "a culture of compassion" to describe the spirit of volunteerism at our school. In fact, his greatest source of pride as dean occurred when our school was awarded the American Association of Medical Colleges' first-ever Community Service Award. All things considered, we hoped we'd find a receptive ear.

Barth started off the meeting with his vision of an international health care program that would be the envy of every other medical school in the country. He had come up with a name for the program—Project Medishare—that captured a spirit of partnership he hoped would develop between us and our Haitian colleagues. Michel detailed how his connections with Haiti and the Haitian-American community in Miami would help. I emphasized the educational opportunities; not only would the students see things they'd never see in the United States, they'd learn a lot about life and about themselves. There were also research possibilities and grant opportunities.

Bernie listened intently, poured over the photos we brought with us, and asked a few questions about the political climate. Finally, it was time for his judgment.

"Sounds great guys. Just stay out of trouble and don't ask me for any money."

With that pronouncement, the meeting ended. Barth, Michel, and I walked together toward Barth's office to regroup. "I was expecting a little more from Bernie," I muttered. "Why should the 'culture of compassion' stop at the water's edge?"

"Relax, Art!" chided Barth. "We got what we needed. It's not an easy time to be a medical school dean, you know. The managed care companies are forcing us to drastically discount our fees, draining our clinical revenues. Everyone is out to sue us, and old-fashioned philanthropy is drying up. We're not ready yet to be on the center stage of the medical school's mission. Besides, we don't need any of the medical school's money. We'll set up Medishare as its own charity. I'll chip in some money to get us started, and we'll raise money through events and from my wealthy friends."

"My brother is a famous Haitian artist," volunteered Michel. "We could have an art auction."

These were credible ideas. Barth, as one of the world's most sought-after spinal cord surgeons, had operated on the back of what seemed like every famous celebrity and athlete in the world. Michel's brother could connect us with other artists and dealers to get paintings donated or purchased at a discounted price.

"Are there any Haitian football players in the National Football League?" Barth asked. He had raised a considerable amount of money for his Project to Cure Paralysis from professional football players.

"We play a different 'futbol' in Haiti," laughed Michel. "You call it soccer."

Nevertheless, we were hopeful that through some combination of benefactors, fundraising events, and eventually grants Project Medishare would take off. I was assigned the task of identifying student volunteers and finding a lawyer who could incorporate Medishare. Meanwhile, many members of the group who went to Haiti on our first trip volunteered to serve on Medishare's board and I was elected Medishare's first president. My Haitian colleagues laughed and called me "Président à Vie" ("President for Life"). You see, in Haiti being president is more of a curse than a blessing. In the entire history of the country, only two presidents had served out their full term. The rest died, were assassinated, or were removed by coup. Being President for Life in Haiti usually meant a short term in office.

One of my first assignments was to try to get some added surgical training for Marlon and Jerry. They wanted a fellowship in gastrointestinal surgery, as there were no surgeons with these special skills in Haiti at the time. Unfortunately, the chair of surgery balked. "How do I know they'll go back to Haiti?" he asked. Medishare had created its first disappointment. Frustrated, Marlon and Jerry found a fellowship on their own, in France. Shortly thereafter, they left Haiti for what ended up being three years.

A few months after our meeting, Bernie retired. The university conducted a search and appointed a new dean, an ophthalmologist by training who rapidly became absorbed in the problems of managing the medical school and the second-largest health center in the nation. Medishare fell off the school's radar screen for a while.

House Calls

1995 "As bad as it is in Port-au-Prince, it's worse in the country-side," Miriam had told us on our first visit. "The people there have nothing—no water, no sanitation, no medicine, no electricity, and no roads." After seeing Cité Soleil, I was skeptical that anything could be worse. But Miriam encouraged me to come out to Pestel and see for myself.

The trip from Port-au-Prince to Les Cayes, approximately 110 miles on a poorly paved road, took four hours. After spending the night in Les Cayes, we embarked in our four-wheel-drive vehicles, equipped with altimeters, inclinometers, and internal/external thermometers, on the 50-mile, five-hour journey across the mountains that form the spine of Haiti's southern peninsula to the coastal village of Pestel. The following morning, Aussibien, a Haitian boat captain, took us in his sailboat to Cayémites, Au Basse, and Zetoit.

Each mile sharpened the contrast between the natural beauty of Haiti and the poverty its people must contend with. The southern peninsula is lush compared to the rest of the country. We saw mountains tumbling into the sea, beaches whose only signs of human presence were abandoned dugout canoes; and forests of palms, vanilla, orchids, and poincairnas. Not far from Port-au-Prince we noticed small houses scattered through the countryside, with families cooking over open charcoal fires. Women washed their clothes while their children bathed in a low-lying aqueduct built by the side of the road. People traveled on foot, on donkeys, or in impossibly crammed multicolored buses. A child walking toward us, carrying a small tin bowl on her head, held out one hand and rubbed her stomach with the other. She was pencil-thin, about eight years old, and dressed in an old tattered party dress. Lynn, who accompanied Miriam and John (Barth's missionary college friend) and me on this trip, was particularly moved by the girl's gesture, so he asked our driver to stop and gave her some of our food. Children in nearby homes who witnessed this act of charity were soon scurrying toward our car, forcing us to hastily speed away.

There was a perfumed scent in the air that mingled with the scent of charcoal, diesel, and cooking oil. "What's that?" I asked our driver.

"Vétivert," he responded, "a perfumed grass that grows in abundance on the southern peninsula. The peasants ship it to France, and it's distilled into perfume." I could make out the large trucks hauling this in front of us, heading for Les Cayes. As we traveled into the interior, the road became rockier and steeper, the houses smaller and more fragile. We passed through Camp Perrin, a small town that looked like it belonged in a western movie, complete with wooden sidewalks and hitching posts in front of its main buildings. We bought some supplies at the general store and then ascended up the mountains.

The altimeters registered 3,000 feet. We pulled off by the side of the road halfway between Les Cayes and Pestel for a "deworming clinic." High in the mountains, people live in huts constructed of thatched and woven palm fronds, connected to each other by footpaths. Miriam was right: There was no water, electricity, or sanitation. Yet the people had their families, their traditions, and their little plots of land—their birthright. The government back in Port-au-Prince could do little to help them, but neither did it oppress them. Mostly, the government seemed invisible way up there. Alongside the crushing poverty, there was pride and hope. Tité called out in Creole to the nearest hut. The mother there sent her four children scurrying toward us and called out to the next hut. Within 10 minutes, *teledyòl* ("word of mouth") had produced 50 children.

All the children had the red hair and swollen bellies of malnutrition. According to the routine, they each got a dose of piperazine syrup (we brought gallons), followed by a chunk of bread and then a piece of candy. I was entrusted with this last task. If the candy were unwrapped, there was no problem. If, however, I gave a child candy in a wrapper, he or she stared at me with a look that asked, "What do I do with this?" I had to teach them how to take the wrapper off before they would eat the candy.

We stopped for lunch in the village of Joli Gilbert. Miriam had set up a small malnutrition station there and owned a piece of land on which she hoped to build a clinic one day. From there we could look all the way down the northern slope of the mountain to the village of Pestel below. We unloaded our supplies at the malnutrition center while the woman who worked there prepared a meal for us in her home on the other side of the road. It was the first home in rural Haiti I had been invited into. The walls were made of wattle and daub, whitewashed inside and out, with blue trim. The roof was thatched palm and the floor hard-packed dirt. The furnishings in the front room (we did not enter the back room, which served as a communal bedroom) were sparse—a homemade table, some homemade ladderback chairs with rush-matted seats, and a charcoal pot over which she was cooking us a lunch of rice, beans, and chicken. Lunch was interrupted by the sound of singing from the road outside. "Oh, it's a wedding!" exclaimed Miriam, evidently recognizing something about the hymns being chanted by the procession. We left our lunch to wish the bride and groom well—she in her formal bridal dress, he in a tuxedo, both riding mules behind a double line of singing, palm-carrying guests.

After lunch we rapidly descended a rocky road that traversed the slope down to Pestel. We wanted to be settled in by nightfall. Arriving just as market day was winding down in the village square, we inched past throngs of people on foot and on mules heading in the opposite direction.

Pestel had a beautiful natural harbor surrounded by a riot of colorful, ramshackle buildings. The town cistern was broken, so even in this large town there was no water. Electricity came on only sporadically, Miriam explained.

The streets in the main square were still clogged with people, engaged in end-of-the-day buying and selling. The harbor was full of dugout canoes. A small island on the other side of the harbor was rimmed with minuscule houses. Six wooden sailboats, in various stages of construction, lined the waterfront. Four recently completed boats lay at anchor.

We rented a canoe to paddle over to the island and inspect the sailboats. They were totally made by hand—their thick planks ripped from logs with handsaws, the ribs carved with axes, the seams of the sails stitched by hand. The spaces between the planks were caulked with rags; then the boats were filled with saltwater, making the wood swell and the boat watertight. The colors—vivid blues, yellows, and greens—were the same as I had seen on the sailboats used by the boat people seeking refuge in Miami 15 years earlier.

Over dinner that night I asked Miriam and the others if this was a place where many Haitians departed for the United States prior to the interdiction policy. "Here and every harbor along the coast," was their reply. "It costs 1,500 U.S. dollars to make one of these boats. That's just about what they sell them for when they're done. The captains are unbelievable. They navigate by the stars and the shape of the mountains on the horizon at night." As we were speaking, I looked over the balcony to the harbor below. Aussibien was finishing his preparations for our journey the next day.

"What time do you want to leave in the morning?" he called in Creole as he cast off.

"Seven," we responded in unison. He disappeared into the dark-

Haiti's coastline.

ness for the hour-long trip down the coast to the village of Zetoit where he lived. We were awakened by the roosters at 4:00 a.m. Aussibien appeared on the horizon at 6:30 and was at our dock at exactly 7:00. His boat had the look of a Chesapeake Bay skipjack, with a gaff-rigged sail and a small outboard motor bolted to the back.

We spent the day making "house calls" on remote islands—Cayémites, Au Basse, and Zetoit, where most people had never seen a doctor in their entire lives. Everywhere, families kept inviting us into their homes to see their loved ones—people with malaria, tuberculosis, and worms diagnosed without x-rays or laboratory tests. AIDS seemed invisible, very far away, something to worry about on a future trip. How providential it seemed that, for no logical reason, we appeared at that time to help so many people we met. But how diabolical that so much time had passed before we arrived and how much more time might pass before our return. And Haiti, although poor, was not the poorest spot in the world or the most isolated. There were other places where conditions were probably worse, where there is nothing and no one to alleviate the suffering. At least this place had Miriam and her organization.

Cayémites was little more than an inhabited dead coral reef and Au Basse an inhabited sandbar. The only vegetation was a fringe of mangroves at the water's edge. Cayémites had 700 families that subsisted by building boats and fishing. The village had one well, actually a cave that dripped water out of its porous sides. An old woman stood at the bottom of the well collecting these drippings into a bucket. Since the bottom of the well was clearly below sea level and the sea was less than 100 yards away, the water she was collecting had to be brackish. There was another small basinlike pit carved out of the coral closer to the sea. This makeshift cistern collected rainwater for the dogs, chickens, and goats of the island to drink. Wiggling mosquito larvae created a geometric design of intersecting waves on the otherwise stagnant water. The husband of the house immediately downwind from the cistern anxiously invited us in to see his wife. She was sweating so profusely that a puddle of sweat formed on the dirt floor between her feet. A few sentences in Creole confirmed the diagnosis of malaria. In addition to giving her chloroquine, we suggested to her husband that he pour some oil onto the cistern for mosquito control.

Au Basse was closer to the mainland and therefore more populous and noticeably more affluent. The homes were larger, and some even had paint. Au Basse had a dispensary, which was the home base of Aldrich, a nurse who served the three communities. Aldrich spent two days a week on each island and then had one day off. He immunized the children, tested for tuberculosis, and arranged consultation with the government-sponsored doctor in Pestel for his most difficult cases. Aldrich's office consisted of an examining table, a chair, and a charcoal pot, over which he cooked us lunch—fish freshly harvested from the sea.

The last patient I saw that day was in the most peripheral home in Zetoit, the furthermost village we visited. The family insisted I see her, claiming she was gravely ill. I entered a small, two-room house with a dirt floor, one door, and no windows. My patient was sitting on the floor in a brown dress, the color of the floor. The dim light made it difficult to see her, except for her eyes and her smile. Although her family referred to her as a child, she was obviously a mature woman. She was retarded, with a small head, and both legs paralyzed. Her family obviously took good care of her. She was clean and well nourished. *Why have they asked me to see her? I thought to myself. She's had this problem from birth. They must think I have magical powers, that I can just lay my hands on her head and restore her to health, that I might be more powerful than a bokar.*

"*Nou pa kapab fè plis. Bon kouraj*" ("We can't do any better than you. Take heart"), I told her family members, who were surrounding me and expecting a miracle I'm sure.

As is their custom, everyone thanked us, whether we helped or not. They fed us and sheltered us in their homes. They gathered at the dock to wave goodbye and sang to us as we departed. And I thanked them in return, though I'm not sure they understood why, as we began our long journey back to Miami. As gratifying as the trip was, it was also troubling. On the one hand, it was a return to the pure joy of being a doctor, unburdened by the bureaucracy and liability and insurance issues that were stultifying my practice in Miami. But I was frustrated by my limited fluency in Creole and by my inability to practice without the bare essentials of medical technology taken for granted at home. And Medishare needed a home of its own, serving a place that no one else was serving, that no one else wanted.

Dèyè Mòn Gen Mòn

1996 Project Medishare had been in operation for two years. Junia stuck her head through the doorway of my office.

"Were you expecting a delegation from Haiti?"

"No. How big a delegation?"

"Just three people. They're from a place called Thomonde. They want to get one of the generators."

Barth had arranged for Jackson Memorial Hospital to donate four huge generators it was recycling to Project Medishare. They were originally intended for the university hospital in Port-au-Prince, but eventually that hospital's administration turned them down— too many issues involving operational costs and maintenance. We decided to offer them to any Haitian entity that could demonstrate need and promised to maintain them.

"They've caught me at a good time. I'm just doing paperwork. Show them in."

Junia left and returned a minute later with the *Komite Zamni Thomonde* (Friends of Thomonde Committee). Of the three, Bob, Mathieu, and Delva, only Delva had actually come directly from Thomonde. The others were Haitian-Americans living in Miami, although originally from the same area. Bob had heard about the generators via *teledyòl* ("word of mouth") and invited Delva to come up and make his pitch. Delva was the magistrate of Thomonde and—as I would come to find out, a visionary—totally dedicated to the well-being of his citizenry. All three wore long-sleeve white shirts, red and black ties, and black trousers. Their demeanor expressed their seriousness of purpose: They were on a mission to earn one of those generators for Thomonde.

Even though Bob and Mathieu spoke better English, after the introductions, Delva did all the talking. He was a short, stocky man who spoke with that perpetual Haitian smile, a smile that reminded me of Régis. Thomonde is more than just a village in the central plateau. It is actually a whole section or "commune," roughly equivalent to a county, which included four other villages and the surrounding countryside, with approximately 50,000 people. Historically, in the eyes of the government in Port-au-Prince, Thomonde did not exist. There had been no electricity in Thomonde for more than 20 years, which adversely affected health and community development. Delva then rattled off a string of statistics documenting just how bad things were in Thomonde and how much good could be accomplished—refrigeration, a hospital, improved education, jobs—if only they had power. There is an expression in Creole: *Lespwa fè viv*, "Hope makes us live." Delva was clearly pinning a lot of his hopes for Thomonde on one of our generators.

"That's very impressive," I told the committee, after Delva finished his presentation, "but it doesn't work like that. It's not within my power to just give you one of these generators. Everyone wants one and everyone claims they need one. You have to write a proposal. It's competitive. The generators will be awarded according to merit."

The committee members conferred among themselves for a minute. "No problem. We will have it to you by tomorrow morning." Sure enough, Delva returned at 8:30 the following morning with a 20-page plan to bring electricity to Thomonde. He asked if I would review it to be sure it was what I wanted. Remarkably, it was

the best proposal I had received. "If I give you one of these generators, Delva, how will I know it is actually being used as you have
written here?"

"Dr. Fournier, you have my word and an invitation to be my
guest in Thomonde whenever you wish. You will stay in my home."

Six months later I received a letter stating that the Thomonde
generator had, indeed, arrived, been set up, and was functioning.
Delva ended by repeating his invitation to visit Thomonde. I
decided to take him up on his offer. The health statistics Delva provided in his application were appalling: 200 cases of active tuberculosis. Eighty-five percent of the children malnourished. No health
care providers or facilities for 50,000 people. Perhaps this was the
"place of our own" Medishare was looking for. In any event, the
place, Thomonde, so isolated and remote, and the person, Delva, so
sincere and determined, intrigued me. How improbable that from
the middle of nowhere with no modern means of communication
with the outside world, he discovered we had generators to give,
found his way to Miami, successfully competed for one, and not
only got it back to Haiti but got it back up to the town, which I was
told was six hours from Port-au-Prince on one of the world's worst
roads. And then he was polite enough to write me, when he could
have just taken the generator, sold it on the black market, and put
the cash in his pocket.

Delva met me at the airport as he had promised. We rented a
4 × 4; stopped at a gas station for ice, water, and supplies; and headed
out for Thomonde. Shortly after leaving Port-au-Prince, a large green
sign marks the point where Route Nationale 3 branches off, thread-
ing through the interior of Haiti to Thomonde, Hinche, and points
beyond. That sign is a perpetual source of humor for me now, as its
large size and block letters complete with a diagonal exit arrow create
the impression that Route Nationale 3 is a real road, perhaps even a
highway. In actuality, shortly after the sign the pavement ends and
the road becomes a dust/dirt/rock/hole/puddle obstacle course that
crosses a desert, three mountain ranges, several rivers (some with

Two visionaries—Dr. Paul Farmer and Delva, Magistrate of Thomonde.

bridges, some without), and a savannah, finally ending in Cap Haitien.

Haiti is a small country about the size of New Jersey. But it deserves to be a country nonetheless, not only for the uniqueness of its language, history, and culture but also for the fact that it physically takes so long to traverse, making it seem larger than it really is. Delva warned me that his six-hour time estimate from Port-au-Prince to Thomonde was just that. The trip, if complicated by breakdowns, mud holes, tap-tap turnovers, or washouts, could take twice as long. Recent improvements have cut the time to three and a half hours, but for the last hour and a half it is still, arguably, the worst road in the world.

Shortly after Croix des Bouquets the road crosses an arid plain carpeted with cacti and thorn acacia trees. The road then ascends rapidly up the face of Mòrn Kabri. This is one of the most desolate regions of Haiti, totally treeless and infamous in Haiti's history as a dumping ground for the victims of the Ton Ton Macoutes. Yet even in its desolation, there was still a steady stream of people—some on foot, some on horse or mule, and a seemingly endless stream of trucks so top heavy with people and produce that each one seemed destined to topple over at every twist and turn of the road. White limestone dust raised by these trucks coated everything—rocks, vegetation, people—along the way and obscured the view of the road as it hugged the side of the mountain. When we reached the summit, we were greeted with a spectacular view of Lac Cayman, La Plaine, Port-au-Prince, and the mountains of the south beyond. The far side of the mountain turned greener as we descended toward the Val Artibonite, and the way of life, as seen from the road, was correspondingly less harsh: men working rice paddies with sticks to plant and hoes to plow, women with straw hats and hand-made livery riding burros to market. Wide deserted stretches of the road inexplicably transformed into impromptu markets, with thousands of vendors and buyers and piles of rich Haitian produce—eggplant, carrots, onions, and mangoes—freshly butchered goats, and still-living-but-doomed chickens and guinea hens. The road followed the river and the continuing flow of Haitian life in the countryside—women washing clothes and people of all ages, but particularly children, swimming and bathing. Around noon the schools let out and suddenly the road was flooded with children, clustered in color-coded uniforms (each school with its own color) walking hand in hand.

Later, we passed through Mirebalais, a fair-sized town for the area, and stopped for a Haitian cola at a roadside restaurant near the cathedral. Above Mirebalais the river valley steepened into a canyon, and the once-tranquil river boiled with cascades and rapids. The road also seemed to boil with ruts that resembled solidified cascades and rapids. After we passed the abandoned cement mixers and the dam that created Lac Péligre (Danger Lake), the road once again climbed the side of one of the mountains that surround the lake. Since the road here was only rocks, there was no dust, and the view was breathtaking—a huge mountain lake surrounded by peaks on three sides, with the Dominican Republic off in the distance. The

View of Lac Péligre from Cange.

lake was dotted with fishermen in dugout canoes. Tiny unpainted wattle and daub houses clung to the hillside between the road and a cliff that tumbled into the lake. Each home had a family outside carrying on their affairs in full view of their neighbors. At the crest of the peak the road left the lakeside and we found ourselves at the small town of Cange, home of Klinik Bon Sauveur. Delva stopped the truck. As we entered the clinic, looking for Dr. Paul, Delva whispered as if we were in church: "Dr. Paul. He's a very good doctor."

When Project Medishare was first organized, I had contacted several other medical schools to find out who had been working in Haiti. Dr. Paul Farmer clearly stood out. He had a combined degree in anthropology and medicine from Harvard and had been working in Haiti for 12 years. He spent six months each year at Klinik Bon Saveur. He had written books about his work in Haiti and used the proceeds to support his clinic. His dean (whom I happened to know through a mutual friend) had told me that Dr. Paul was the only living saint he had ever known. His professional life—indeed, even his personal life—reflected total commitment to what he called "a preferential option for the poor," a belief that poor people had the same right to health as the rich and shouldn't get second-class care.

Klinik Bon Saveur sat on a promontory overlooking Lac Péligre. Its architecture was half functioning hospital and half gothic monastery. "Leave it to Harvard to nail down the prime real estate," I joked, to break the ice. Paul launched into an apology (in the philosophic sense) about the clinic and the people it served. The people in the village we saw as we drove up were all squatters. They had been flooded out of their ancestral homes when the Duvalier government built the dam that created Lac Péligre. Built with USAID dollars, the dam was intended to generate hydroelectric power for Port-au-Prince. Of course, that never happened: No one ever figured out how to keep the intake valves clear.

"So that's progress in Haiti," deadpanned Paul. "The power never worked and the people lost their land. There is a lot of poverty here in Haiti, but that's the difference between decent poverty and indecent poverty. Land is the Haitian peasant's birthright. For years, the United States was afraid Haiti would go communist, so the government poured millions into Haiti to prop up the Duvalier regime. In the process, thousands of Haitians were displaced from their land. No longer able to subsist and support their families, they slipped into the squatter state you saw as you approached." I made a mental note not to joke with Paul ever again.

I asked Paul how we might work together. Paul is all about service and commitment to the poor. He wasn't interested in volunteers who showed up occasionally with no thought about follow-up or continuity of care. He was intrigued, however, about a long-term partnership for Thomonde between Medishare and his charity, *Zanmi Lasante*. Thomonde was his biggest problem. Without access to health care, the people in Thomonde were close enough to Cange to come when desperately ill, but not close enough to come for routine care. Thomonde's population dwarfed that of Cange. Paul's sta-

A traditional home in Thomonde.

tistics amply documented the problem—those estimated 200 cases of active tuberculosis in Thomonde were only one example. So it would please Paul greatly if some group like Medishare would "adopt" Thomonde and work with Delva and the people there to develop a permanent health system. As he walked us back to our truck after the meeting ended he gave me a gentle admonition. "Just don't be like so many others Art, who've come, made promises, and never come back. If you do that, you'll break the people's hearts."

Thomonde was 15 kilometers or an hour's ride after the crest of Cange. As we rounded a ridge, I could see that Thomonde lay at the bottom of a huge volcanic caldera: a green oasis amid the grasslands and deforestation of the central plateau. The main village of Thomonde was a pick-up-stix scattering of streets and cross streets, each lined with traditional Haitian two-room homes—a living room and a communal bedroom. Most were painted bright turquoise blue with white trim. A large mapou tree, decorated with Voodoo symbols, marked the first fork in the road. Delva's house was of similar design but totally white, with a large tournelle—a large round open space covered by a roof—that served as a village gathering place.

Thomonde, like Mirebalais and its neighbor, Hinche, appeared on colonial maps of Haiti. The town was originally settled by the Spanish in 1630. They were probably pushed out by "maroons" (runaway slaves) in the 1700s. The name Thomonde was derived from the Spanish *todo el mundo*. (all the world). The reason to give this name to a town so isolated escaped me. Somehow, Todo El Mundo changed into Thomonde, roughly translated as "Tom's world." It was built on the French village model, with a town square in front of a church at the center of town. Delva took pride in the town square, a small park and a gazebo that he built with money from the European Union. He introduced me to the village notables and then asked me to address the village via megaphone from the gazebo. I met the vice-mayors, the schoolteachers, and Bernardeau, a local success story who had emigrated to Boston years earlier and sold used cars. Now he was wealthy by Thomonde's standards and returned to Thomonde every winter to avoid Boston's chill. A party erupted with dancing and rum and clapping. Project Medishare, in the person of Dr. Fournier, had come to Thomonde.

Delva gave up his bed and home and moved in with a neighbor for the night. I marveled at the workmanship of the shutters and the doors of Delva's house. When the kerosene in the lamp burned out, I was left in the darkness with my thoughts. Voodoo drums pounded in the distance. Their rhythm and intensity helped me focus. Here was a place no one else wanted but a place that wanted us, a place where Project Medishare could make a difference.

The roosters started crowing at 2:30, coupled with the yelping of dogs and the braying of mules. At 5:00, the church bell started chiming and shortly thereafter the farmers started out for their fields. At 7:30, as the schoolchildren marched to their schools, I took a walk around the town. The children whispered and giggled as I passed. Some of them silently followed me. One, a natural leader even at six years old, took my hand and smiled with silent pride, the envy of all her peers.

Delva came by with breakfast around 8:00—wonderful Haitian coffee, fried plantains, and a Haitian style of spaghetti. It was the beginning of a beautiful friendship. How could he know that my mom is Italian-American and that pasta for breakfast is my ultimate fantasy? We sat under his tournelle in the coolness of the morning and he told me of his dreams for Thomonde. Health care was at the top of his wish list, even higher than electricity. "If you don't have health, you don't have anything. Here, if you don't have your health you can't work. If you can't work, not only do you die, but your children die also." I thought about how many people I know who take their health for granted or, worse, abuse themselves. "You're very wise, Delva," I responded, "for a politician."

Later, we took a walk to the old dispensary, which had been closed for three years and was in a sad state of disrepair. "This is the only health care facility in the entire commune," Delva confided. We then took our 4 × 4 to a field outside town. There, in a little shed, stood the generator. "The town of Thomonde owns this land," Delva whispered conspiratorially. "Wouldn't it be a great location for a hospital?"

"Someday, Delva, perhaps, but we have a saying in English, 'You've got to crawl before you can walk.' I can come back in two months and bring a team of medical students with me. We can only stay for a week. It's not much, but it's a start."

Before we left Thomonde to return to Port-au-Prince, Delva took me to make a house call on his parents. They lived in a small house on the edge of town, with a surprisingly large plot of land, neatly demarcated by cactus hedge rows. His father had suffered a stroke in the recent past, and his mother had some pain in her left shoulder that caused Delva some concern.

Before I examined them, Delva introduced me and pulled over a handmade ladderback chair for me to sit on. His mother had been cooking soup for her husband over a charcoal fire outside. She was short and, like Delva, always smiling. Instead of the usual head wrap, she wore a cheery blue broad-rimmed bonnet. She beamed with pride as I told them the story of the generator and Delva's first trip to Miami. There wasn't much I could do for his father. The stroke had crippled his right side and made it extremely difficult for him to speak. The smell of urine permeated his bedclothes. I suggested he take an aspirin a day and that the family try to get him out of bed as much as possible. His mother had a simple case of tendonitis. I gave her an anti-inflammatory medicine that I had in my suitcase. In half an hour her pain was gone—Medishare's first miracle in Thomonde.

Tom's World

Six weeks later . . . "*TI POUL, DELVA! TI POUL!*" cried Susan S. from the passenger side of our rented truck. She clutched the dashboard as if it were a brake lever. "*Pa gen pwoblèm,*" chuckled Delva as he swerved the truck to avoid the latest mother hen and chicks to dare cross our path as we bounded and crunched our way from Pignon to Thomonde. The swerve lifted us out of the rut that traffic had worn in the dirt road. Although Delva was accommodating to Susan's desire not to kill any chickens, he was not going to slow down. We had to get to Thomonde. There would be lots of patients waiting.

"*Ti poul-yo,*" I corrected Susan's Creole. "There is more than one chicken." It had become a running joke now: Susan's "I brake for animals" vigilance balanced by Delva's determination to complete the four-hour road trip from Thomonde to Pignon to Thomonde by 10:00 a.m. Every time we bounced on a boulder the students in the truck bay would scream "Yahoo!" Children watched us cautiously from behind their cactus hedge rows or ran laughing after us yelling, "*Gade blan! Gade blan!*" ("Look at the strangers!")

Seeing "*blan-yo*" in this part of Haiti, deep in the interior, close to the Dominican border, eight miles from anywhere, was certainly an event. I had kept my promise to Delva to come back with a team of medical students to do a health fair in Thomonde. By screening for common, preventable, treatable problems, such as worms and malnutrition, we could start the process of returning health to Thomonde. Project Medishare was not an official part of the University of Miami's curriculum, but word was passing from student to student that there was a real opportunity to learn in Haiti. As a consequence, more and more students were giving up their spring, summer, and winter breaks and volunteering for Medishare. On this trip I had 12 second-year medical students, long on book knowledge but short on experience. I was the only real doctor. But that was the wonderful thing about the health fair method: One doctor aided by a dozen students could see a lot of patients.

Delva showed me the Uzi he kept under his seat. As magistrate of Thomonde he carried absolute authority, and he wanted us to know that he personally guaranteed our safety while visiting Thomonde. The students were impressed with the Uzi, particularly since the *Miami Herald* had written another series of articles about gang violence and police ineffectiveness in Port-au-Prince just before our trip. Delva was more concerned about reports that there were barricades blocking the road around Hinche. The report turned out to be nothing more than rumor. Neither barricades nor gangs ever materialized. I wasn't worried. In my 20 trips to Haiti in Medishare's first two years of existence, I had learned that there were really two Haitis—the political Haiti you read about in the newspaper and the real Haiti hardly anyone outside the country ever sees. In the media's version of Haiti, the country was in a perpetual state of political violence and crime. In the real Haiti, at the time, political violence was frequently more symbolic than real, a kind of political theater, usually confined to the capital. Crime in the countryside in those days was practically nonexistent. In Haitian culture the worst thing you could possibly be is a thief, a *vole*. People policed themselves. And if anyone should seriously break the rules, there was always the threat of the zombie curse.

Certainly, I had seen some petty theft in Port-au-Prince. There was one street corner near the airport where youth gangs looked for tourists in unlocked cars, opened the doors on the run, and tried to steal purses or jewelry. My Haitian friends said these were second-generation Haitian-American youths who grew up in New York City, were corrupted by America, and were then deported after criminal convictions. "They don't even know Creole," I was told. Even with this pocket of petty crime, Port-au-Prince seemed much safer than Miami. And in contrast to the capital, there was simply no crime in Thomonde. Neither Delva nor the people would tolerate it, and in Thomonde there would be no place for *vole* to hide. Besides, as Dr. Paul had put it, the poverty was "decent" in Thomonde. Every peasant proudly worked his own little plot of land, compared to the "indecent" poverty of the uprooted masses in Port-au-Prince. During my first visit to Haiti, I was amazed at the peace and security I sensed in Cité Soleil. Now the politics were so unstable that my Haitian friends advised me not to take any students in there. Fortunately, Medishare now had Thomonde.

There were only a few AIDS cases identified in the commune of Thomonde at that time, probably because it was so isolated in the interior and because there were no doctors to diagnose it. The low numbers would surely grow. There were many cases down the road in Cange and even more in Mirebalais. There was more extreme poverty there, increasing the number of people traveling to and from the capital looking for work, forcing more women to turn to prostitution and more men to leave their families to cut cane in other parts of Haiti or in other countries, particularly the Dominican Republic. As tortuous as the drive from Port-au-Prince had been, the road is the only link between the countryside and the capital. In Haiti the AIDS epidemic had evolved in a classic pattern, first appearing in the slums of Haiti's ports and tourist centers and then slowly spreading to the countryside. Paul had already demonstrated how the road facilitated this spread of disease. The two vectors? The police (if you were a Haitian peasant, you couldn't say no to the police in the 1980s) and the tap-tap drivers.

A Thomonde *granmoun*. Photo by Wassim Serhan.

I was skeptical about the estimates I had been given by Delva and Paul about the number of tuberculosis cases in Thomonde. As we were driving around the commune, however, I could visibly identify people with telltale signs of consumption. From the perspective of the victims and their neighbors, it probably made no difference whether they were suffering from AIDS or from the more ubiquitous tuberculosis. I quickly learned that Western medical diagnoses mattered little there. All that mattered was that the people were visibly marked for death.

We pulled off the road and into the field that surrounded the newly finished magistrate's office. Hundreds of people who had gathered outside cheered our arrival at exactly 10:00 a.m. The day before we had seen about 200 children—plotted their heights and weights, identified those who were malnourished, and treated the ones with worms and impetigo. We had promised the *granmoun* (the adults) that we would see them today, but first we had to transport the sickest children to the hospital in Pignon, about two hours away. Some of the students stayed behind to give health education classes on malnutrition, oral rehydration, family planning, and HIV prevention.

"*N'ap kòmanse nan senk minit. Mèsi pou pasyans ou!*" ("We start in five minutes. Thanks for your patience!"), I announced as I pushed through the crowd and past the iron gate entrance to our makeshift clinic. "How'd the classes go?" I asked the students gathered in the central foyer.

"Great! You could tell by the questions the patients asked that they were really getting it. Jean-Gason and Jean-Peter [two boys from the local school who were fluent in English and had volunteered to translate] did a super job!" volunteered Susan A.

"But Dr. F. We've got a problem."

"What's that?"

"All those people want to be seen! We counted nearly 600."

"Well, I guess we'd better see them. We've got 12 students, so that means two students in each exam room, plus two to guard the gate and regulate patient flow, and two to man the dispensary. Organize yourselves, and let's get started."

The crowd was already pressing against the iron gate, but the people backed away just enough for me to exit and announce, "*Granmoun premye!*" ("Old folks first!"). In Haitian culture old people are revered. *Granmoun*, which means "old person," also means "wise person." Adhering to this tradition, we decided to see the oldest patients first. The sea of people in front of the gate parted as four men carried in the oldest citizen of Thomonde on a palette and helped her into an easy chair.

"*Bonjou, dokte,*" she said as she was carried by me.

"*Bonjou, mami,*" I responded. "Susan, Felix—would you please take this first patient? Use Jean Peter to help you."

Remarkably, in 10 minutes the clinic was in high gear. The students performed limited histories and physical examinations, while I rotated from room to room, identifying the problems and how we'd treat them. Then the patients would go to the "dispensary" (a table set up in the central foyer with the essential medicines we had brought from Miami) to pick up their medicine. The only problem was that there was only one gate out, and everyone was trying to get in.

"Dr. F., This first patient says she's 115 years old. Could that be right?

"It's probably pretty close, since her daughter who's with her tells me she's 84."

"What do you want us to do for her?"

"Find out why she can't walk. Do a complete exam. Don't worry, take your time."

After an hour it was clear that the patient's complaints could be grouped into certain general types of problems—*bouton, gratél,* and other skin conditions; heart problems; fevers; masses and hernias. Though the makeshift clinic was running pretty smoothly, we had a major problem: How to get the patients we had already seen out without letting the increasingly anxious crowd in.

"We've finished our exam and don't find anything terribly wrong," said Susan A. "She's just a little unsteady on her feet," referring to the 115-year-old woman. I stopped my running from room to room when it hit me.

"That's the oldest person I've ever met . . . and probably ever will!" I went over to talk to her—a sweet woman who looked remarkably like her daughter. She would like to walk but was afraid of falling, she said.

"We've got this walker we carried from Miami. Do you think that would help?" volunteered Felix.

"Great idea!"

Five minutes later the 115-year-old woman who was carried in walked out under her own power. The crowd roared its approval. This seeming miracle, though, only increased each person's desire to be seen, so the crowd pushed even harder to get through the gate.

The windows of the magistrate's office were open, with the hands and faces of children pushing through. One of the hands gave me a note. It was from a child with asthma we saw the day before. I wanted to see him back for a recheck and had instructed him to meet me at the front gate. I went out through the crowd and found him and led him in by the hand. The crowd, if anything, was getting bigger.

"Dr. F, we've reached gridlock. No one can leave. They've pressed so tightly against the gate, so we can't bring in anyone new," exclaimed David, one of the gate's guards. "I'm a little worried someone might get hurt."

"I think we've got a revolution of rising expectations on our hands," I muttered and then attempted crowd control to no avail: *"Pa pouse, pa pouse!"* ("Don't push!") David and Tom were leaning all of their weight against the gate to keep it shut. Finally, the hinges gave way and the 500 citizens of Thomonde still waiting to be seen poured through the gate and down the two corridors of the magistrate's office.

"All of you, stop seeing patients! Gather up your things! We need to get out of here!" I screamed as I withdrew to the central foyer.

In five minutes, as the last of the patients streamed in, the students and I were able to march out, single file, carrying our precious medicines in duffel bags. I announced in Creole that people needed to leave the magistrate's office immediately or we would never return to Thomonde again.

It seemed like I was the only one who took the situation seriously. The patients were laughing and the students were smiling, amazed by the spectacle of it all, as we retreated toward the truck. A circus atmosphere prevailed. Delva, who had been working on making us lunch, came running over from his house, looking worried. *"Kisa k'ap rive?"* ("What's going on?"), he asked me as he caught his breath.

"Yon ti revolisyon, se tou" ("Just a little revolution, that's all"), I responded. Delva's voice obviously carried more authority than mine. *"Sòti kounye-a!"* ("Get out now!"), he barked, and in minutes his office was empty again. He told the patients we were going to take a

lunch break and reorganize, that we'd return and would attempt to see everyone, once we had established a reasonable plan.

Delva whisked us away in the truck to a secret place—a relative's house—where we huddled to see how we could solve the patient flow problem. We returned to Delva's office by 3:00. We had the patients form four lines—one for skin problems, one for heart/blood pressure, one for fevers, and one for masses and hernias. These groupings were important. Although it's true that few people died of skin problems, the misery they caused was extreme and they were easy to treat, which would build confidence among our patients in Western medicine. The fever line was critical. With a few simple questions we could sort out patients with malaria, which we could treat on the spot, from those with suspected tuberculosis, whom we'd have to send en masse to Dr. Paul in Cange. The hernia line might seem trivial, but with the only surgeon in the region two hours away in Pignon, I knew that we'd find a lot that had never been repaired. In Thomonde a hernia could keep you from working, and if you couldn't work, you couldn't feed your family. Each student team was assigned to a single line, with each line forming outside a single room. We were back in business again, each line moving at a noticeable pace, and with no pushing, or pressing.

"What are we going to do with all these hernias?" asked Ian, one of the two students in the hernia room.

"We'll take the four worst ones with us to the hospital in Pignon tomorrow. To the rest, we'll explain, that if it gets stuck and won't go back inside, it's an emergency and they'll need to go to Pignon. Otherwise, we'll try to bring a surgeon back and fix some more when we return in July."

Miraculously, by 10:00 p.m. we had seen all the patients in each line, and the students were finishing up on the last patients in each of the exam rooms. We were almost out of medicines. While I was physically exhausted, my mind was in overdrive from the stimulation of seeing 600 patients. I was proud that we had kept our promise to see every patient, and I told the students so, but now it was time to quit. The generator that Delva had rigged to give us light would run out of diesel fuel soon. I told a few curiosity seekers on the front steps that we had to stop. We needed to be up at 5:00 a.m. to take the patients with hernias to Pignon.

"Dr. Fournier, you need to see this kid," said Tom as he approached me from behind.

"We saw kids yesterday, Tom, and it's time to quit. . . ." My sentence trailed off as I turned to face Tom, holding a limp infant in his arms.

"His mom carried him from way out in the countryside. He's had diarrhea for two or three days and he's not nursing well," explained Tom, sounding almost apologetic after my rebuke.

A quick exam on the front steps while Tom held the infant in his arms revealed a limp, sluggish, and difficult to arouse four- or five-month-old baby. The sunken soft spot in his skull, dry mouth, and wrinkled skin made the diagnosis of life-threatening dehydration obvious.

"Sorry, Tom. You're right. Take this child and his mom inside. Mix up some oral rehydration therapy. Do you remember the formula? Two liters of water, a half cup of sugar, and a pinch of salt."

Tom (holding the baby), and the mother ran up the steps to one of the exam rooms. I tried to treat my own dehydration with one of the Haitian colas provided by Delva. "What's next?" I asked myself as the cool soda soothed the back of my throat. After coaching 12 medical students through the care of 800 patients over two days, I needed a break. As if on cue, Tom appeared.

"Uh, Dr. F. We've mixed this stuff up, but how do we give it?"

Of course, there were no baby bottles in Thomonde! I walked over to the exam room where the child was sprawled limply on the bench. By now the students had gathered round, all looking to me for a solution. Momentarily stumped and straining for an answer, I explained the signs of dehydration that the child exhibited and how dehydration from simple diarrhea was the major cause of death in children in Haiti. In my mind, however, I was searching for an answer to the immediate problem.

A straw, with force feeding? We had no straws. The finger of a latex examining glove stretched over a cola bottle and pricked with a needle to form an artificial nipple? No, the latex was too thin for even an infant to suck on. We had already distributed all of our condoms.

Then, suddenly. "Get a syringe, Tom, and take the needle off the hub." Tom ran to the dispensary and brought back one of our few remaining syringes. I filled the syringe with oral dehydration solution and offered the hub end to the baby's mouth while I applied gentle pressure to the plunger. The infant's lips and tongue instinctively wrapped around the hub, and a few seconds later the child swallowed. It was going to work! "O.K., students, we're going to rehydrate this infant one cc at a time. Each of you take turns doing exactly as I'm doing; then we need to teach his mom how to do it herself. She'll have to get him through the night. If he makes it, we'll take him up to Pignon with us tomorrow. I'll ask Delva if he can find a home they can stay in tonight."

The child's mother was a quick learner, and in what seemed like no time the child had consumed 30 cc's of the solution. Already, he seemed more active and more responsive. I sent her off into the darkness with Delva and a two liter bottle of the rehydration solution, with instructions to give the baby as much as he wanted and to be back at 5:00 a.m. for a ride up to Pignon. I told the students to clean up a bit—we had really trashed Delva's office—and then gave the order to douse the generator. We plunged into darkness and climbed into our cots or sleeping bags.

Despite my exhaustion, I had difficulty falling asleep. Voodoo drums pounded in the distance. In the gray zone between sleep and consciousness, the first few minutes of hearing voices in Creole and English made me think I was hallucinating. I forced myself to focus on where the voices were coming from—outside on the steps. I

A mother in Thomonde with a large number of malnourished children.

looked at my watch on the chair next to me but could not see it in the dark. I got up and lit a candle: 2:00 a.m. I pulled on my shorts and stumbled toward the gate. Jerome (the troubadour of Thomonde, Jean called him) and two of his friends were talking earnestly with two of my students about music and philosophy. Jerome had stopped by earlier to entertain us, but we were still seeing patients. Evidently, the students were having trouble falling asleep also. I met Jerome on my first visit to Thomonde. Delva arranged for him to play for me and my students. Not only does Jerome sing in French, English, Spanish, and Creole, but he also writes his own songs. He's lived his entire life in Thomonde, except for an occasional visit to Port-au-Prince. In my mind, he's the perfect allegory for Haiti—so unique, so talented, so unjustly unknown. Jerome played his guitar and his three Haitian friends harmonized on a simple folk song about a young Haitian who goes to the Dominican Republic to cut sugar cane to earn money for his family and whose wife leaves him for another man while he's gone. We finally broke up at 3:00 a.m. When I awoke 90 minutes later, I was surprised at the early morning activity in Thomonde. People were already cleaning their stoops, washing clothes, or filling water buckets at the village spigot across the path from Delva's office. I wanted to be ready to go when the students, who were due to be up at 5:00, arose, so there I stood at the spigot, washing myself and sharing my soap with the early-rising Thomondois.

The flatbed truck that Delva had arranged to take us to Pignon arrived precisely at 5:00, along with the four people with hernias and the mother, the infant, and an almost-empty bottle of rehydration solution. The child was so much better—alert, aware, squirming in his mother's arms, that I almost considered not taking them to Pignon. On the other hand, a couple of days of observation to make sure the child didn't relapse wouldn't hurt. His mother had dressed him up in a small suit and booties. The students trickled out of their rooms, threw their gear in the truck, and climbed aboard. Mother and child were sitting on the spare tire, along with Tom, who wanted to be close to his "save." The other students stood, holding on to the side bars and crossbars of the truck as we jerked into gear and headed up the road. I waved to my bathing mates at the spigot as we left. "*Abyento!*" Jerome's song from the night before kept running through my head. It stayed with me all the way to Pignon.

Exposition Santé

THAT WAS OUR FIRST HEALTH FAIR in Thomonde. We've subsequently returned, about three times per year, with an ever-growing cadre of committed students. The early health fairs were pretty chaotic. Michel and I were the most consistent trip leaders. In a community that had received health care only episodically from missionaries, we had to break through an unspoken communal mind-set in which every patient pushed to be seen, shared a litany of complaints, and hoped to walk away with some magic pill, any pill. It didn't matter whether it would really help them or not.

In the beginning, malnutrition was rampant in Thomonde. It was easy to screen for: Simply line up all the children with red hair. Hair is, in essence, pure protein. The first signs of protein deficiency show in hair. The hair of malnourished children in Haiti changes from curly, thick, and black to thin, sparse, and red. Human beings are 95 percent water. One of the many miraculous things that protein does is to keep water in our cells and circulatory system, rather than us dissolving into an amorphous puddle. As protein deficiency progresses, therefore, a child's belly and feet swell with fluid. Malnutrition is further complicated by worms. When we first went to Thomonde, we just assumed each child had worms. Therefore, each health fair in the early days always had a "deworming line." The red hair and swollen bellies of malnutrition were not just clinical signs of the disease; they were signs for all to see that these children were somehow marked for misfortune and probably premature death. While in Ireland red hair might be considered beautiful, in Haiti it's a symbol of shame.

One family was particularly poignant—two devoted parents and five daughters, all under the age of 5, all slowly starving to death. I explained to my students how the mother had breast-fed each child but had to stop prematurely to return to her work selling produce in the market. With this many children, none could be adequately nourished. As I was telling the mother what kinds of foods she should be giving her children, she burst into tears. Her children were marked by a particularly unusual sign of malnutrition: Their hair turned not red but blonde, and she blamed herself. Worse, there was nothing she could do about it, as the family was too poor even to raise chickens, let alone goats or a pig. I was ashamed of my insensitivity. Before I left them, I touched her hand and whispered "*kenbe-la*" ("hang in there").

Riding back to Port-au-Prince, one of the students, Stephanie, volunteered that her husband worked for a company, Rexall-Sundown, that made vitamins and protein supplements. She would write a letter to the president of the company, requesting a donation. Stephanie's letter resulted in a donation of 14 palettes (approximately two tons) of multivitamins and protein supplements. With the help of a Peace Corps volunteer, we started feeding and prenatal care programs. Shortly thereafter, the kids in Thomonde started having black hair again.

Over time our health fairs became more organized. The focus shifted from acute care to screening and prevention. I think of Tim at the start of every health fair. He'd be surprised and delighted to see the unexpected way I fulfilled his dying wish. If he could have held on until HAART (Highly Active Anti-Retroviral Therapy) arrived, he'd probably be in Haiti with us.

We started going out to the remote corners of the commune—
Baille Touribe, Tierre Muscadet, and Savannette. It takes as long to
get to Baille Touribe from Thomonde as it takes to get from
Thomonde to Port-au-Prince, and the road—built by hand by the
Thomondois and Baille Touribois—is even worse than Route
Nationale 3. It's hard to imagine a more isolated corner of the planet
than Baille Touribe. Totally cut off from everything but Thomonde
and Cange, not even within range of the nearest radio station, it's
surprising the Baille Touribois have not evolved their own language.
With no electricity, cars, anything remotely modern, the people live
in a timeless, traditional way. Most Baille Touribois rarely travel more

A family of four, Baille Touribe.

than 10 kilometers from the place of their birth during their entire lifetime. We began doing our health fairs in the church in Bas Touribe, a cool shady place, with a good breeze when its doors are open. The priest let us use his vestibule—a small room off the altar—for Pap smear screening.

Pap smears had never been done in this part of Haiti—a Project Medishare first. We found a lot of positives, but imagine the difficulty of explaining the need for and the process of such a procedure to people living in such a timeless, isolated place. So we started with a class in Creole outside the vestibule door that began with the basics: What is a cell? What is cancer? Does anyone know someone who died of cancer? Hundreds of women lined up and patiently waited all day for their turn to climb up on the makeshift examining table and submit to the examination. Pap smears continue there today.

While the students set up for the screening, I usually circulate among the crowd, explaining that we're not there to treat every ache and pain but to screen for problems that might kill them. I also scan the throngs for the obviously ill who need to receive individual attention. They're easy to identify with their wasted frames and gaunt, frightened stares. The patients I can visually identify as having AIDS or tuberculosis invariably are standing off by themselves, shunned by the crowds. Failing health can only be explained by a curse, you see. Both the victims and their neighbors understand that.

The health fairs proved to be invaluable for screening and prevention. They were also a clever way to engage and organize our enthusiastic but inexperienced medical students. Since Medishare had first come to Thomonde, we had brought the commune from no health care to episodic care, with screening and patient education. Episodic care was certainly better than no care at all, but the visiting American doctors and students were never going to break through the stigma and shame of AIDS, tuberculosis, and malnutrition. We'd need a permanent presence. And it would have to be Haitian.

Close Encounters

1998 MY BEEPER WENT OFF WHILE I was giving expert witness testimony at a malpractice trial in West Palm Beach. Annoyed, I pressed the button to stifle the beep without interrupting my testimony. I rechecked my beeper during recess. The number started with an area code I was not familiar with. "Must be a mistake," I remember thinking, but if I'm compulsive about one thing, it's answering my pages. I dialed the number on my cell phone.

"Open Society Institute. Farzi speaking. How may I help you?"

"Hi. This is Dr. Arthur Fournier in Miami, Florida. Did you page me?"

(Farzi chuckled softly) "This is the Open Society Institute, New York City. Just one moment."

"Hi. My name is Ellen. Congratulations. We've decided to fund your project, 'Family Medicine Training in Rural Haiti.'"

"Wow! Thanks! Um, is this official? Is there anything else I need to know?"

"I'll be writing you with all the details. It's a challenge grant. That means we're only giving you half of what you need. Another foundation will need to match our challenge, but we'll help you find that partner. The First Lady will be announcing the award in Haiti

185

when she visits on November 22. That's in three weeks. Can you be there?"

"Can I bring Michel, my co-conspirator?"

"Surely. We'll work through the details. She'll announce it at a ceremony at Dr. Guy's hospital in Pignon."

"I'm giving testimony at trial right now. This is for real, right? Is there a number where I can call you back tomorrow?"

"Just use this same number. The First Lady wanted to announce something positive while she was in Haiti, so she called us to see if we had anything in the works. We reviewed all the proposals and yours was the best."

I had just enough time to call Michel before the trial resumed. After our first trip to Haiti, it was clear that what we could do that would help Haiti the most was what we also did in Miami—train family physicians. The Haitian medical education system, modeled after the French, had only one residency position for every two graduates. These residencies in traditional specialties were all based in Port-au-Prince. Thus, the system forced its graduates to either specialize or emigrate. Meanwhile, in rural Haiti, where 85 percent of the people lived, there were hardly any doctors. Those who did practice outside the capital were not trained to meet the health care needs of the people they were supposed to be serving. We put these ideas down on paper, organized into an ambitious program to fund three residency programs throughout the country—one in the north, one in the center, and one in the south. Now, at least one was going to become a reality.

Michel and I had to book reservations in first class in order to get to Haiti with only three weeks' notice. We drank champagne and congratulated each other during the flight. Mostly, we were thrilled that our plans for training Haitian doctors in family medicine were about to become a reality. But this scholarly achievement was about to be infused with star quality. We were going to meet the billionaire philanthropist George Soros and the First Lady, Hilary Rodham Clinton. We rendezvoused with Mr. Soros, the philanthropist who founded the Open Society Institute and Ellen at the Hotel Olaffson, a classic Caribbean gingerbread design hotel in Port-au-Prince. George (he insisted we call him George) had made his fortune in international finance. In financial circles he's known as the "man who broke the Bank of England," the ultimate capitalist. But he gives most of his money to his charity, which is dedicated to transforming countries formerly ruled by dictators into democratic, open societies. Casually dressed and unassuming, he bought us drinks at the bar. "Now tell me, what's family medicine, and why is it so important for Haiti?"

Most Americans don't realize that the discipline of family medicine is a modern American invention. In fact, it was invented in large part in Miami in the mid-1960s, by Lynn, my roommate during our first trip to Haiti. Our medical school created the first department of family medicine and the three-year residency training program in family medicine. Prior to that, doctors going into general practice did only a one-year rotating internship. Best suited for rural communities, family medicine had not taken hold in some U.S. urban enclaves, such as Boston and New York. Michel and I were not surprised, therefore, with the tone of Mr. Soros's question.

Michel reviewed for him the theory behind family medicine— the need for broad-based generalists with special skills in primary care to deal with 90 percent of the health problems that affect poor Haitians, with special skills in primary care. Cleverly, he explained how, in a country like Haiti, the family medicine model complimented his vision of an open society. The profession of medicine in Haiti was, until now, a closed society. Even though medical education was free, only the sons and daughters of the elite could afford the education needed to pass the test for admission to medical school. Most graduates chose an urban specialty private practice after graduation or emigrated to France or the United States. Objectively, a poor country like Haiti simply could not afford the luxury of one doctor for its children, one for its adults, and one for its pregnant women. George listened intently.

I had been a little left of center since my college days. But after an hour of some of the most stimulating conversation of my life, I found myself thinking, "Gee, maybe capitalism isn't so bad after all." I promised George and Ellen that of all their grants they would never get as great a return on their investment as they would from our training program.

That evening we attended an informal reception at the U.S. ambassador's residence. After short speeches by President Preval and Mrs. Clinton, the audience broke into small groups on the lawn for cocktails while the ambassador, the first ladies, the president, and Mr. Soros went inside for a formal dinner. We spent the night at Michel's mother's house in Petionville and then headed for the general airport for the charter flight to Pignon.

Pignon boasts "the third-best airport in Haiti." It was built to accommodate doctors visiting Dr. Guy's Hôpital Bienfaisance. It would be hard to imagine a more difficult hospital to get to if it were not for the airport—six hours from Port-au-Prince and three hours from Cap Haitien, on the world's worst road (the same one that passes through Thomonde). Not a problem for the First Lady and her entourage; they'd come by helicopter, a half-hour flight.

Pignon is a dusty village on the northern edge of Haiti's Plateau Central. Dr. Guy's father was Pignon's pastor, the only formally educated person in the village. Growing up, Guy would visit the sick with his father. From these visits grew his resolve to be a doctor. Somehow, after his medical school in Haiti and residency in surgery in New York, he became a flight surgeon for the U.S. Air Force, and he used his savings to fund his hospital. We had visited it as Project Medishare volunteers and deemed it perfect for family medicine training.

When our chartered Cessna circled the field and landed, we were the first dignitaries to arrive. Both Michel and I wore dark suits and ties, as dictated by Haitian protocol. I was envious of the children running around naked, as the temperature was at least 90 degrees. A small crowd gathered at the airport building—equally divided among

First Lady Hilary Rodham Clinton and Mr. George Soros announcing the grant for Family Medicine training.

village dignitaries anxiously awaiting the First Lady and the curious, sure something important was happening but not sure exactly what it was.

Protocol officers from the First Lady's office, Secret Service agents, and officials from the U.S. Agency for International Development scurried about trying to organize the unorganizable. They wanted a rope strung on poles to separate the people from the First Lady. Unfortunately, none of them spoke Creole and not a single Pignonois spoke English. So I translated and soon a 30-foot cordon extended from both sides of the airport building. To the people, this rope was meaningless. They wandered around both sides or out onto the runway, following their cattle or goats. Michel found a tarantula on the grass runway and dutifully notified the Secret Service. They called in a government 4 × 4 vehicle which, in the interest of security, ran over the poor spider six times.

With the field somewhat secured, we waited in anticipation of the First Lady. Like the climax of the movie *Close Encounters*, two small helicopters and one large one appeared first on the horizon and then hovered slowly toward the runway. The crowd cheered, and the local high school band played enthusiastically. Aliens arriving, I thought, but friendly aliens. Once on the ground the First Lady and her guests, including George, the world's second-richest man, and Bill Sr., father of the world's richest man, disembarked. George was the guy actually giving us the money. Bill Sr. was literally along for the ride, learning the philanthropy ropes. The group was greeted by Dr. Guy and the mayor of Pignon, both with red sashes, and two charming little girls in yellow and pink dresses carrying flowers. The entourage piled into their SUVs and we headed for the hospital. Family medicine training was about to arrive in Haiti.

The First Lady wore a pink pantsuit. After we returned to Miami and developed our pictures, this was the number one topic for discussion, with the world divided into two camps—those that loved it and those that hated it. She toured the hospital with Dr. Guy, while Michel, myself, George, Bill Sr., Dr. Guy's parents, the Open Society Institute entourage, the protocol officers, the Secret Service agents, the hospital workers, and a sizable portion of the population of Pignon waited in the courtyard. To her credit, Mrs. Clinton spent extra time talking to the cadre of *fanm saj*, the midwives trained by Dr. Guy, who was the principal reason why the hospital had reduced its maternal mortality by two-thirds. Each proudly carried her metal box containing sterile scalpels, ligatures, and gauze pads.

Guy, George, and Mrs. Clinton all spoke eloquently, and George introduced Michel and me. We stood and bowed to the cheers of the hospital employees and villagers. Bill Sr. was fidgeting, concerned that he hadn't gotten to ride in the First Lady's helicopter. Between speeches Ellen darted between his seat and hers. After she promised him he could ride with the First Lady on the return trip, his mood lightened considerably.

A young woman strode to the podium, also dressed in pink. She spoke passionately about the difference a micro-loan had made in her life. She had bought a sewing machine and now was supporting her family. Confident, assured, the Haitian peasant eclipsed the billionaire and the First Lady. I glanced at Michel. How strange life is—what accidents create the rich and the poor, and is it justice or injustice that united the seamstress and the president's wife? More importantly, the young woman's impassioned speech convinced me that we had to think beyond health and start thinking about ways, such as micro-loans, that would address the underlying cause of most of the country's problems—Haiti's grinding poverty.

Afterward, I asked Ellen if it might be possible to meet the First Lady. "Sure, no problem! Just wait by the hospital entryway. We'll have to pass through to get back to our cars." It turned out that Ellen and Mrs. Clinton had been friends since college.

"You know," I said to Michel as we waited for our big moment, "it was really nice for Mrs. Clinton to come here and do this. I mean, how many votes is she going to pick up by coming to Pignon?"

Just then, Ellen, Mrs. Clinton, her photographer, and the Secret Service entourage turned the corner. As promised, Ellen stopped and graciously introduced us.

"Thanks so much for caring about Haiti, Mrs. Clinton," I gushed, starstruck as I shook her hand.

"Call me Hillary, please. And don't thank me. It's you two who are doing the work!"

Marasa

Of all the villages in rural Haiti, Labadie would seem the most improbable for Project Medishare to visit. The tiny village is nestled between Morne Cap and Labadie Bay, prosperous by Haitian standards. Most of the adults are employed by Royal Caribbean Cruise Line, a Miami company that owned all the property surrounding the village. Labadie's pristine beaches were among the cruise line's most popular destinations. The villagers maintained the property, manned the concession stands, ferried passengers between the cruise ships and the beach, and cleaned up after they left. The cruise line referred to Labadie as its "private island," never mentioning that it was in Haiti.

We were there for the most pragmatic of reasons. Royal Caribbean Cruise Line had offered to make a substantial donation to Medishare if we'd do a health fair in Labadie. For all its prosperity, there was no real health care in Labadie, and health issues were taking a toll on the productivity of Labadie's workers.

To celebrate our coming, the villagers decided to spruce up the abandoned government dispensary, located on the edge of Labadie Bay with a view of the tourist beaches and plantation ruins beyond. They had stocked the pharmacy and hired a handyman to plaster the most obvious cracks and apply a fresh coat of paint. About half an hour before the health fair was scheduled to start, Joseph, my lead student, called me over to see the handyman. "I think he's got typhoid," Joseph said.

One glance confirmed Joseph's diagnosis. The handyman was lying on the dispensary floor, sweating profusely and clutching his stomach in pain. Next to him was a foul-smelling pot of fish stew.

"Gather up the rest of the students, Joseph. We've got our first patient."

I asked the students to find some Cipro, mix up some oral rehydration therapy, and start giving the antibiotic and the fluid to the patient. After they started treatment, I explained: "One picture is worth a thousand words. See that fish pot? We don't see typhoid in the states anymore because of things we all take for granted—clean water, sanitation, and refrigeration. Here in Haiti, even in prosperous Labadie, there is no refrigeration. So our poor patient brought a pot of fish stew with him and tried to nurse it through his two days of work here. It's a good thing we came when we did. He could have died. Joseph, check his vital signs every half hour and make sure he keeps up on his fluids. Everyone, let's pick him up gently—his abdomen's exquisitely tender—and move him to the back room, out of the way. We've got to get started. The line of patients already extends to the center of town."

Joseph and I were the only bilingual members of the team. Joseph, being a Haitian-American second-year student was an invaluable resource. I had enormous confidence in him, as I knew him well. As a first-year student, he had been assigned to me in the clinical skills course. So he was my protégé. I actually didn't have to teach him that much. He had intuitive people skills. He spoke French, Creole, and Spanish, fluently. He had only started learning English when he emigrated to the United States at the age of 14, so his English was marred by a heavy accent. When stressed, he stammered to find the right words, but it took a lot to stress him.

During Joseph's first year, I had him work with Phil, whose

Achilles' heel as our best doctor and teacher was that he could not speak Spanish or Creole. At the time, we had a patient from Honduras with a wasting syndrome strongly suggestive of AIDS. The patient was in denial, however, and refused to be tested. It was Joseph who not only convinced him that he should be tested but also explained the results in Spanish to the patient and his wife. The couple had been separated for several years, he having been forced to return to Honduras by the Immigration and Naturalization Service, but now that he had returned and was ill, she was his only source of support. At first she was so upset to discover her husband had AIDS that she threatened divorce. Joseph's people skills won the day, though. He not only convinced the wife to be tested (she also tested positive) but counseled them to both start therapy and stay together. They are alive and well to this day.

Joseph was a proverbial gentle giant. He towered over me and showed his affection for me by rubbing my head, as if it were a bottle with a genie in it. I'd become something of a father figure to him, in part because of my devotion to Haiti and in part because his own father had died when he was younger. He grew up in St. Marc, a dusty ramshackle colonial town on the coast an hour north of Port-au-Prince. He worked three jobs while in his preclinical years to pay his tuition, but still found time to volunteer for Medishare. This was our third trip together.

Joseph took each patient's history as he registered them, told it to a student who took the patient into an exam room, and reported the results to me. From there the patient went to the pharmacy for medicines and then out the door.

Soon the health fair was humming. As we got busier and busier, the improbably beautiful setting of the clinic by the sea faded from my consciousness. This could be a health fair in Thomonde or anywhere else in rural Haiti—lots of children with malnutrition, worms, and scabies. By noon the handyman felt well enough to return to his painting. "Make sure he's got a week's worth of Cipro and knows how to take it," I told the student manning the pharmacy.

"Hey, Dr. Fournier. Help me with this exam!" called Ana from the prenatal station. "I can't be sure of this baby's position." Part of good prenatal care is to determine the baby's position in the womb. Normally the head should be down. A breech-first baby almost always requires a Caesarian delivery. Not knowing in advance that you need a Caesarian is pretty much a death sentence in rural Haiti. I examined the eight-month pregnant woman carefully.

"Feel here Ana," I said as I placed her hands on the patient's abdomen. "What do you feel?"

"Oh, so that's where the head is!"

"Now feel here."

"But that feels like the head also!"

"Now listen here with your stethoscope."

"Oh, the baby's heartbeat!"

"Now listen here."

"Another heartbeat? Dr. Fournier, I'm so confused!"

"What's your diagnosis, Ana?"

Ana looked at me dumbfounded.

"*Marasa!*" I told the patient with a smile.

"Twins!"

"Omigod, Dr. Fournier! You can do that? Diagnose twins without an ultrasound?"

Maybe I was better prepared to practice medicine in Haiti than I realized. One of my pet peeves about how medicine was changing in the United States was its increasing dependence on technology. In fact, I was firmly convinced we were being enslaved by it, unable to practice without it. So diagnosing twins the old-fashioned way gave me particular pleasure.

Twins are special in Haitian culture, and this particular mother-to-be left very happy. Joseph, however, had found another set of twins, and they weren't so fortunate. "Look at this, Dr. Fournier. The twins were identical, as evidenced by a photo the healthy one carried with him that had been taken a few years before. In the interim, one had developed extreme weight loss, chronic cough, and

Identical twin brothers, one with AIDS, one without.

diarrhea. His physical exam showed the telltale yeast infection of AIDS. The healthy brother looked about 30. The sick brother looked three times older. His healthy twin implored us to help.

"He's probably got SIDA [the initials that stand for AIDS in French] and tuberculosis," I explained in Creole. "The best we can do is send him to Hôpital Justinien au Cap. I know the doctors there. They'll take good care of him."

Joseph picked up the patient from the examining table and, with the patient's arms around his neck, carried him out of the clinic. He waded into the bay, hoisted him over the gunwales of an anchored water taxi, and hailed the captain.

"You'd better go with him," I suggested. "We're almost done here. Look for one of the family medicine residents and make sure he gets admitted." I gave him $20 for the water taxi and helped the brother climb in. *"Bon dye beni nou!"* ("God bless you both!"), I called as the boat turned and headed into the setting sun.

The next morning I had a meeting with a village elder, Franco. I expressed my concern over the quantity and severity of health problems I had seen the day before.

"There's not much we can do," said Franco. "Our only source of water is the stream that flows through the village. We have no electricity and no doctor."

"I'll talk with the people at Royal Caribbean when I return, and we'll see if they'll help. You seem to have an AIDS problem here. Where did that come from?"

"It came from the path."

"The path?"

"Yes, the path. I'll show you on your way out."

Off the trail that connects two beaches in the tourist part of Labadie is a small but well-worn footpath marked by some painted rocks. The path leads up to a small clearing in the woods that cover the mountainside behind the beach. According to Franco, it's there that the village girls and boys wait for the tourists on Mondays and Wednesdays and sell themselves for $10 or 10 euros. Tourists interested in casual anonymous sex know about the path or find out about it shortly after their arrival. Somehow, they slip away from their spouses or partners or group; trudge up the path with their snorkel masks, sunscreen, and flippers; and enjoy the pleasures of the harbor. The stream is pretty constant, from the first disembarkation until the last whistle call. The exchange rate doesn't seem to matter. Smart virus. Smarter than we are.

Mambo

MEANWHILE, THE UNIVERSITY OF Florida School of Medicine, Miami's state-supported sister school, was following in Project Medishare's footsteps. Nestled in the small north Florida college town of Gainesville, UF seemed worlds apart from Miami. Surprisingly, though, it always had a fair number of Haitian-American students from Miami in attendance—students willing to give up the comfort and support of home in exchange for Florida's much lower tuition. One of them was Serge.

Serge somehow found me at Miriam's orphanage, where a small Medishare team was performing physical exams on the kids. He had been home for spring break visiting his family in Carrefour, on the other side of Port-au-Prince. He had heard via *teledyòl* ("word of mouth") that some *dokte blan-yo* ("foreign doctors") were working at an orphanage in Post Cazeau. He and his brother, Rousseau, borrowed the family car, an old red Pinto, drove across town and, in a manner similar to my first failed attempt to find Père Luc, asked and searched, asked and searched, until they found us.

"Hi! My name is Serge. I'm a first-year student at the University of Florida School of Medicine. Are you Dr. Fournier? I heard about what you're doing from my cousin in Fort Lauderdale. She works as a nurse at Jackson. There's a group of us at UF trying to organize to work here in Haiti, but we don't have any faculty. Do you think you can help?"

Serge was tall, very dark, and somewhat reminiscent of a young Sidney Poitier, except with a squarer jaw. He had a heavy accent and spoke with a little bit of a lisp. That day he wore a T-shirt and jeans. I acknowledged that I was Dr. Fournier and invited him to join in with our exams. I then peppered him with questions, in part to get him involved and in part to assure myself that he was, in fact, a medical student. He passed the test.

As it turned out, the University of Florida's spring break always occurred the week following Miami's. It was therefore relatively easy for me to stay on an extra week each spring, supervise my Miami students, and rendezvous with the Florida students at the airport in order to supervise them. Serge had returned to Gainesville after our first encounter and organized his fellow students to work in a similar manner to Medishare. Unlike the University of Miami teams, however, Serge and his team set their own itinerary. The first two trips on which I worked with UF students focused on orphanages in Port-au-Prince.

"So, where are we going, this time, Serge?" I asked during a phone call in anticipation of our third trip together.

"We're going to Jacmel. We'll be doing health fairs in villages right behind the beach at Raymond-le-Bain. But first I've found a new orphanage that needs our help. It's up in the mountains on the way to Jacmel in the village of Decouze. They have about a hundred kids. There's only one problem, Dr. F."

"What's that Serge?"

"It's a Voodoo orphanage."

I think Serge expected me to be shocked. "So? We've got a Catholic orphanage [Luc's] and a Protestant orphanage [Miriam's]. This will make us truly ecumenical!"

About 20 miles out of Port-au-Prince, the road to Jacmel branches off the road that snakes down Haiti's southern peninsula, ascending and then descending over the peninsula's mountainous spine. In contrast to most of the roads in Haiti, the road to Jacmel was newly paved—Jacmel had an influential legislator who hoped a paved road would increase tourism. So even though the road twisted and turned as it climbed the mountainous ridges, it was a relief from the dust and ruts of Route Nationale 2. The view as we ascended was spectacular—first the plain of Leogane, then off in the distance Isle de la Gonave and the mountains of the southern peninsula, tumbling into the sea. These mountains were greener than those to the north of Port-au-Prince. Nestled between the Caribbean and the Golfe de la Gonave, they get more rain. As a result, the terraces carved out of the side of the mountains seemed more productive. The peasants farming there seemed relatively prosperous, with larger plots of land, more distance between their neighbors, slightly larger homes, and much more impressive tombs.

By this time I'd been to Haiti often enough to know that everything there has meaning. I tried to explain this to the Florida students as they took in the homes, markets, farms, and cemeteries we passed along the road.

"Let's start with the tombs," I began. "In New Orleans, they tell the tourists the tombs are built above ground because the city is below sea level. Really, the above-ground tomb is a Creole tradition, seen all around the Caribbean. It's part of Voodoo. An elaborate tomb shows respect for deceased ancestors. Look there! You can see an offering left of grain and rum."

"That tomb is more elaborate than the house next to it!" exclaimed a student, pointing through the window of the van to a three-tiered crypt with a gothic facade.

"That's because for a Haitian your tomb is your home for eternity. You'll be spending more time there than in your earthly home," explained Serge.

"And notice the colors and the crosses. Blue and white are Christian colors, black and red are Voodoo. Then there's a cross for the

Catholics and a different-shaped cross for each Protestant denomination."

"Is that why all the houses are painted blue and white?" exclaimed another student. "I just thought they must have had a paint sale for those colors."

"No, it's the same symbolism," I continued. "If you see a red and black house, it's probably a *ounmfò*, a Voodoo temple. And notice the carvings on the wooden eaves. They're all symbols of either saints or voodoo spirits."

The road dipped and weaved through Tombe Gateau ("Fallen Cake," the mountain on the other side of the ravine looked exactly like that!) and then ascended to Decouze, the highest point on the mountain ridge. The students' emotions alternated between awe at the spectacle of Haitian life unfolding before us, the beauty of the terrain, and the terror of oncoming trucks whipping around hairpin turns. As we approached Decouze, a mist (or was it clouds?) intermittently surrounded us and the temperature cooled noticeably. When the mist cleared, we could see both the Caribbean and the Golfe de la Gonave at the same time.

Up ahead the road was clogged with burros and women selling wares, a traditional Haitian market. Fortunately, we had reached our destination—Orphelinat Bon Secours. The name was painted stylishly in large black script on a rose-colored wall that surrounded the whole compound.

"How did you find out about this place, Serge?" I asked. "Do they know we're coming?"

Serge explained that the orphanage (actually it was both an orphanage and a school, and some children from Decouze with parents were allowed in for classes during the day) was founded and financed by a bookstore owner Serge had met in Port-au-Prince. His wife, Madame Felice, not only ran the school and orphanage but also served as Decouze's *mambo*, a Voodoo priestess. She was assisted by her two daughters and a nurse. The founder told Serge he would send a message to the orphanage that we were coming via the nurse, who traveled back and forth between Port-au-Prince every weekend. We took the presence of a nurse as a good omen—a willingness of the *mambo* to "work with both hands," as the Creole expression goes—to accept Western medicine as well as herbal and faith healing.

The large red iron gate rolled back in response to our honk and then closed behind us, pushed by two older children. The compound, which filled a bowl-shaped natural amphitheater at the crest of a ridge, was composed of a two-story concrete house painted white, two red barracks-like buildings (obviously the girls' and boys' dorms), a building that served as the school, and another—painted red and black, with a huge iconlike mural facing the courtyard—that was clearly a *ounmfò*, a Voodoo temple.

The ancient Romans put realistic faces on their idealized statues. Similarly, the mural was simultaneously an icon of the Virgin Mary/Erzilie Freda and a realistic likeness of what I assumed to be Madame Felice—a pudgy, smiling face, with gray hair covered with a red and purple veil and a gold or jeweled ring on each finger. Before we got out of the van, I explained to the students the significance of Erzilie, a voodoo goddess available for intercession in matters of love and family. Our hostess was there on the wall for all in the village to behold, the embodiment of the Virgin Mary, Aphrodite, and Athena, all rolled into one.

"She must be some *mambo!*" one of the students whispered.

The courtyard was filled with naked children, washing themselves with water drawn from several large barrels. There was also one nearly naked adult—Jeanne, the nurse—wrapped only in a towel.

"*Ou te rive bonè*" ("You arrived early"), she stated matter of factly and then excused herself to get dressed. "I'll be back in five minutes. Your students can set up in the school. But don't let them get started until you've met Madame Felice."

A few words are in order here about Haitian dress (and undress). Clothes in Haiti are not for modesty. It's a hot country and most people grow up sharing a common room with their parents and siblings. Everyone knows what everyone else looks like. Children, especially boys, run around naked until puberty, and public bathing both at communal wells and in rivers and streams is common for both adults and children. At the time of the Haitian revolution (1792–1804), the fashion statement from Paris was *dezabiye*—partial undress—a custom that continues in Haiti to this day, with dresses open in the back and breasts frequently exposed. Today clothing in Haiti is about status and artistry. The same child running naked an hour before will don a three-piece suit or ruffled dress to go to church or to visit the doctor. While peasants work the fields in hand-me-downs and rags, every family has at least one set of dress-up clothes, cleaned and neatly pressed, for each member of the family, set aside for special occasions. It came as no surprise, therefore, when Jeanne emerged from her room five minutes later in a tailored skirt, jacket, and medium high-heeled shoes. "Follow me," she said, in Creole, as she escorted us into what I assumed was Madame Felice's living room.

The room was dominated by a large painting of a white stallion galloping into the foreground. Several other smaller pictures represented saints and voodoo spirits. A sofa and recliner had probably seen their best days in Miami. There weren't enough seats for all of us, so most of the students sat on the floor. A few wandered down the corridor down to the classroom. Jeanne announced she would return with Madame Felice in a few minutes.

Twenty minutes passed. Hezi, a third-year student from Miami assigned to me for his family medicine clerkship (and the only Miami student with me on this trip) came down the corridor and sat down beside me. "Dr. F., I think there's a kid with appendicitis in the classroom. He's got fever and a really tender abdomen. Did we bring a surgical tray?"

"In the U.S. it would be appendicitis," I responded. "Here it's probably typhoid. We seem to find at least one case every trip. I'll see him as soon as we meet Madame Felice." Just then Jeanne returned.

"Madame Felice extends her apologies. She's not feeling well today. But she will entertain a small entourage in her boudoir." I suggested that Serge and Evelyn, the one faculty member from the University of Florida that Serge had been able to recruit, should join me. I then instructed the rest of the students to set up for physicals in the classrooms.

Serge, Evelyn, and I ascended the stairway to Madame Felicè's bedroom. She was a large woman, dressed in a red togalike robe, being groomed by her daughters, radiating her Erzelie personna. I knelt and kissed the largest ring on her right hand and then introduced the others and explained what we hoped to accomplish—a physical for each child, a test for tuberculosis, and a medical record.

"Do you have any medicines for sugar?" asked Madame Felice. "I have sugar in my blood, and it's impossible to get medicine for it."

I sent Serge back to the van in search of medicine for diabetes. He returned with several boxes, which we gave as a donation to the *mambo*. Pleased, she released us to return to the students and our work. "Jeanne will give you a little tour, but tell your students not to enter the *ounmfò*. It's sacred and off-limits to all except the congregation."

I skipped the tour and returned to see the child Hezi thought had appendicitis. Gaby was 12. He was not an orphan, but rather a student—one of the best in the school. His mother, a market woman, paid a small amount each month to Madame Felicè for his studies. His jet black skin and angular features suggested to me what Régis must have looked like when he was a child. I wondered if there was anything like reincarnation in Voodoo. Gaby was stoic, refusing to show pain. Two touches to his abdomen—boardlike rigid and tender to the slightest touch—confirmed not only typhoid, but typhoid complicated by spread of infection to the lining of the abdominal cavity.

"Give him some ampicillin [the only antibiotic we had] and stay with him, Hezi. You're his nurse! Get some fluids in him. Evelyn, can you help Hezi?" Evelyn was a pediatric intensive care specialist in Gainesville. If Gaby had presented there rather than in Decouze, he certainly would have had the full-court press-monitors, IVs, etc. Here, it was basically a pill and a prayer. Evelyn looked bit like a fish out of water.

The children, freshly scrubbed and dressed now in their finest, waited patiently in line. The Florida students worked efficiently. Before I knew it, it was 2:00 in the afternoon. Evelyn came over as I was checking a young boy for a hernia. "I'm worried about this boy with typhoid. He's not getting better."

"We'll have to take him with us to the hospital in Jacmel," I told her and then asked Jeanne to send a child to the market to find Gaby's mother.

It took about half an hour to retrieve Gaby's mother from the market. She was short but quite muscular, dressed in a white peasant dress, a white head scarf, and sandals. I explained the situation in Creole as simply as I could: Her son had a life-threatening infection and needed medicines we did not have with us. We'd have to take him to Jacmel to save his life. Gaby's mother resisted. She was sure Madame Felice could cure him. She had done so with every other illness he'd ever had. She knew of other children who went to Jacmel and never came back. And then there was the issue of cost.

I told her that some illnesses were so bad that we needed to work with both hands to beat them. This was one of those illnesses. If Madame Felice and she prayed and we got Gaby the medicines he needed, we could probably pull him through. Either approach by itself might fail, but together we should be able to save him. I also promised that we'd pay his bills.

She looked at her son and his obvious anguish and agreed on two conditions: She and Jeanne had to accompany us to Jacmel, and we'd need to give her a small donation she could pass on to Madame Felice for her prayers. *"Dakò!"* I exclaimed as I pressed her hands.

We finished up the health fair and gently loaded Gaby into the back of the open cab truck we had used to haul our medicines and bags. He was cushioned from the metal base of the nine-inch cab by a series of student volunteers' laps acting as cushions, with Jeanne cradling his head. It was getting close to 5:00 and Gaby's mom had disappeared.

"Where's Gaby's mom?" I asked Serge.

"She had to go home. She said he needed a clean shirt and a new pair of shoes!"

No matter that he was too weak to stand. Haitian custom dictated that he couldn't go to the hospital without a clean shirt and decent shoes! The students took advantage of the down time to wander off and peek inside the *ounmfò*.

Gaby was kept in Hôpital St. Michel for three days. The Haitian doctors there pumped him full of three antibiotics at once. He made a dramatic recovery. I visited him every morning, got a hug from his mom, and gestured to her with my hands together, signaling that our Voodoo/Western medicine approach was working. On the day of his discharge, she pulled me aside and showed me his bills for the hospital, the doctor, and the pharmacy. Paying the bills worried me. Three days in the hospital in Miami would cost tens of thousands of dollars. But paying Gaby's bill was literally the price we had to pay to save his life, so I figured, if worse comes to worse, Evelyn and I would phone home and wire transfer the money. I added up the three bills and divided by 17, the conversion rate from Haitian currency, gourdes, to dollars. "That can't be right," I thought, as I checked my math again and again. One hundred gourdes each for the doctor and the hospital and 225 gourdes for the pharmacy—that's 425 gourdes—25 bucks! I laughed out loud, which surprised and puzzled Gaby's mom. I reached for my wallet, took out the correct amount of gourdes, and asked her to check to see if I'd given her the right amount. She kissed me on the cheek and ran to the cashier's window. So life is cheap in Haiti, but saving lives is even cheaper. Gabuy's life cost $30 total, if you include the $5 we gave to Madame Felice for her prayers.

Baptism

THE NEXT DAY JACMEL, ON HAITI'S southern coast, seemed vaguely reminiscent of the French Quarter in New Orleans, with its French colonial architecture, wrought-iron balconies and above-ground tombs. Actually, in my view, there's no comparison. Jacmel, founded in 1697, was a bustling colonial port when New Orleans was a backwater settlement. Nestled between the mountains and the Caribbean and interspersed with bougainvillea and flowering trees, Jacmel looks much as it must have 200 years ago. Gaby had been admitted on Friday. I visited him each day to check on his progress. Saturday we did a health fair on the beach at Raymond-le-Bain. On Monday we would have to return to Orphelinat Bon Secours to read the skin tests for tuberculosis we had placed on Friday (it takes three days for a reaction to develop) and to bring Gaby and his mom home. Project Medishare's teams rarely work on Sundays, and this group had earned a day off, with their hard work in Decouze and Raymond-le-Bain. I was feeling ebullient. Gaby's was a life saved. Whatever else happened, we could take that to the bank. I promised the students that later I'd take them to Bassin Bleu—a 100-foot waterfall and series of connected pools across the river and down the lip of an extinct volcanic crater just to the west of Jacmel. The pools get their name from the curious milk-turquoise color of the water. I told the students, inventing myth as I went along, that the pools are sacred and that if they dive into the pools they are officially Haitian. They were either too naive, too in awe, or too polite to challenge my story.

Being an early riser, I set out on my own at sunrise that Sunday morning to see if I could find a walking route to Bassin Bleu. The usual route required a four-wheel-drive vehicle, a river crossing, several guides, much clamoring for tips, and sometimes begging from women and children along the path. But the topography suggested that the stream should flow down to the sea at the end of the black sand beach that extended a mile to the west, on the other side of the river from the city.

Heading west from the hotel, I passed first through a residential neighborhood. People were up already—children running around *ni net*, gentlemen engaged in earnest discussions, women cooking *akasan* and *avwàn*. The scents of vanilla, cinnamon, and charcoal invigorated my steps. Next I passed through a tiny park on the edge of the sea, populated by young people engaged in walking meditations or committing their *devwa* (homework) assignments to memory in the shade of magnificent palms. On reaching the beach I removed my sandals and felt the unique feel of black volcanic sand under my feet.

Leaving Jacmel I had to first pass a squatters' settlement at the point where the Grande Riviere du Sud empties into the bay. The squatters' homes were perilously assembled like houses of cards on a plain of thick black mud. Pigs were running through the mud, as were children. I had to wade through a small fork in the river to continue on the beach, the mud-stained water coursing black, but cool over the black sand. It being Sunday there were signs of both industry and repose at the river—children meticulously gathering sand shrimp (a delicacy) with a strainer; men loading polished rocks washed down from the river into baskets to be used for construction; men and women bathed while others, their clothes spotless and shoes

Children on the beach at Jacmel, costumed for Carnival.

shined, walked in from the countryside heading to church in Jacmel. The water in the fork was about two feet deep, and the churchgoers stopped at the water's edge, removed their clothes, placed them neatly on top of their heads, waded the stream, and then redressed on the other side. Of course, everyone greeted each other with a genuine "*Bonjou*," even saluting the "*blan*," who curiously seemed heading in the wrong direction. Several people stopped me to ask me if I was lost and offered to guide me back to the hotel.

If Jacmel seemed locked in an 18th-century maturational arrest, the countryside west of the river appeared not only further removed in time but also place. Clusters of small houses formed villages set in among the palms just inland from the beach with curls of blue smoke rising above the thatched huts from the morning cooking fires. Men with handmade hoes tended their fields. Others herded their oxen or gathered in their sein-nets from the beach. Children darted down the paths from the village to catch a glimpse of me as I passed, some stopping at a safe distance, others boldly coming up to me, taking my fingers, and enticing me to play with them. "I could just as well be in West Africa," I thought.

Four fishermen were struggling to launch their boat from the high-water mark. "*Blan, ede-nou souple*" ("Stranger, help us, please"), said one, matter of factly. I put down my sandals and pitched in at the stern. By rocking and shoving and grunting in unison, we, imperceptibly at first and then with increasing successive gains, forced the barque over the rise and then down the slope of the beach to the point where the rising waves suddenly made it seem weightless. "*Mèsi anpil, blan*" ("Thanks a lot, stranger"), said my new-found workmates as they shoved off and I continued west. The black sand abraded the soles of my feet, and each wave left them tingling. The farther west I went, the fewer people I encountered. The beach ended abruptly at the base of a cliff, and, just as I imagined, I found there the little stream that assuredly led to Bassin Bleu. A footpath followed the stream and crossed from side to side, gently climbing the small gorge. The stream carved through an extinct volcanic crater. The path was shaded by *mapou* and royal palms and flowering trees I could not identify. The sweet scent of flowers was everywhere; I felt like I was hiking through the Garden of Eden.

Pulling myself up over a small ledge, I came upon a young woman washing her clothes in a small cascade. "*Bonjou, cheri!*" I called as I approached.

"*Bonjou, mesye,*" she sang. She coquettishly lowered her dress, exposing her breasts.

She looked to be in her early 30s, although she was probably younger. Life in rural Haiti ages one prematurely. She was admittedly attractive, of medium build and muscular with breasts that were relatively large and upright. In fact, I suspected she was nursing an infant. Her laundry pile suggested she had a large family, with children of various ages.

"*Eske ou vle mwen benyen ou, mesye?*" ("Do you want me to bathe you, sir?")

The sense of flattery caused by this come-on was rapidly negated by the knowledge that I was middle-aged and, as a *blan*, ludicrously ugly. *She must be from the village tourists pass through on the way to Bassin Bleu,* I thought. I had found out about Bassin Blue several trips before this one, from Michel, a "*gid blan*" who was always hanging around the hotel. As likable as he was, there was something about the way he would ask, "Dr. Fournier, is there anything you want me to show you? Anything you want? Anything?" that made me wonder if he was strictly in just the guide business. Since the United Nations intervention, Jacmel had become a popular destination for army troops on weekend leave and there were lots of *gid blan* like Michel willing to show the soldiers the sights and more. So the thought of sexual tourism in Jacmel in general, and Bassin Blue in particular, had crossed my mind in the abstract already. But this was real and obviously not her first attempt. I was being seduced and propositioned. My principles were being put to the test. There was no one else around. No one would ever know.

"*Pa pa lage kò-w bay touris pou lajan. Se blan yo ki pote SIDA ak lòt maladi yo. Genyen fyète aysisyen-w. Pa mande!*" ("Don't give yourself to tourists for money! Strangers have AIDS and other maladies! Take pride that you're Haitian and don't beg!"), I said.

She was stunned to hear this mini sermon. "*Ou fou, blan*" ("You are a crazy foreigner"), she said as she laughed.

"*Pa fou, men kontan fè konesans avèk ou. Ou vle danse?*" ("Not crazy, but happy to meet you. Would you like to dance?"), I came back, as I shuffled my hips and feet in a poor imitation of *compa*. At this she pulled up her dress, gathered her laundry, and quickly descended down the path convinced, I'm sure, that I was crazy.

I continued up the stream, past the abandoned house of the Canadian who mysteriously arrived one day, built his home, married a village woman, and just as mysteriously disappeared. I climbed the approach to Bassin Bleu using first the makeshift ladder left there for the tourists and then the handholds and the footholds carved into the rocks by the Arawaks. I had the cascades, the waterfall, the pools, and the sheer rockface covered with ferns and orchids all to myself. I slipped out of my clothes and dove in.

Kay Medishare

2000 I WAS SURPRISED TO HEAR DELVA'S voice on the other end of the phone one Friday afternoon: "Dr. Fournier, can I come over to talk? I have an idea I want to speak with you about."

"I'm really busy this afternoon, Delva. Are you staying through the weekend? I can see you first thing on Monday."

"I'll be there." I could almost see him smiling on the other end of the phone. All weekend I wondered what his idea might be.

Monday morning Delva was there in his white shirt, black pants, and red tie. I knew a pitch was coming. We made small talk—a municipal election was coming up in Thomonde. Delva was probably the most popular mayor in Thomonde's history, but there was a problem—he was not Lavalas, Titid's party. I told him I couldn't imagine him losing.

"We'll see," he answered, as his perpetual smile faded a bit.

The smile came back in full form, however, when I asked him about his idea.

"Dr. Fournier, I have this plot of land in Thomonde. Good land. I want to give it to Medishare in the name of the people of Thomonde, and I want you to build a guesthouse in Thomonde."

"A guesthouse? That's the idea?"

"Yes, that way you'll always have a place to stay when you and your students come to Thomonde. I can't promise you will always be able to stay in the *mairie* (the magistrate's office). Besides, the mairie has only one bathroom and no showers. I want to build you the best guesthouse in the Plateau Central, three bathrooms and three showers."

Despite my outward skepticism, I could see there was a germ of a good idea here. Our health fair trips were growing to sometimes 20 to 30 people and myself and other faculty were also coming in between the larger trips. Medishare needed a home in Haiti. Why not Thomonde?

"How much will it cost me?"

"$25,000. I can have it done in six months."

I had a small discretionary fund I had accumulated by giving my opinion on malpractice cases to attorneys. It would just about cover the cost. "Where else but Haiti can you build a three-bedroom, three-bathroom house for $25,000?" I mused. "Delva, I'm not going to be able to give you an answer today. I've got to talk about it with the Medishare board. When are you leaving?"

Delva was leaving on Wednesday, which gave me two days to poll the board. I started with Michel, whose office was just three doors down from mine.

"Um, Michel. Delva was just here, up from Thomonde. He wants us to build a guesthouse in Thomonde on some land he'll donate to us."

"What, are you crazy?" Michel's jaw fell in disbelief. "Do you know how many poor Haitians there are concocting wild get-rich-quick schemes right now to pitch to wealthy Americans?"

"Delva's not like that, Michel, you know him. It's all about helping Thomonde. I trust him. Besides, I made a vow a long time ago to never again doubt a Haitian. Aren't you getting tired of sleeping on cots, washing out of buckets, and pooping in latrines?"

"That's a good point. What will it cost us and how will we pay for it?"

"$25,000. I've got just about that much saved up in the 'legal account.'"

"Well, if it's your money, it's your call. Just make sure you see the deed and title."

"Of course, I'm not totally crazy, but speaking of titles, think about it, Medishare means that much to him. He's willing to give up part of his birthright."

The other Medishare board members pretty much followed Michel's pattern—incredulity, skepticism, and, finally, resignation. In the final analysis it was my money. Plus it made sense. It was Delva's way of showing not just what Medishare meant to him but also what it meant to all the citizens of Thomonde; a chance for health. And in a tangible way it would solidify the relationship between Medishare and Thomonde. Medishare and Thomonde had been dating seriously. Now, Delva was proposing marriage and putting up his land as a dowry.

On Wednesday, Delva called me from a pay phone at the airport. "It's a go, Delva. But I've got to see the deed and an official title transfer. What about an architect?"

"Don't worry, Dr. Fournier. I'll send everything to you in a few weeks!"

"I'll bring half of the money when I come down in two months and the rest when it's finished."

"I'll see you then!"

Two weeks later I received a letter from Delva with the deed, title transfer, and preliminary drawings, which I was sure Delva had drawn himself. The design was clearly multifunctional: two bedrooms in the back and a third bedroom in a separate wing for privacy—Delva had penciled in "Dr. Fournier's suite." The front was designed to serve as a combination living room and dispensary, with little windows that could serve as pass-throughs for medicine. A large tournelle (circular pointed roof) provided protection from the rain to a patio with a platform stage, which could serve as outdoor living space, a classroom, or for health fairs. In addition to the three bathrooms, the house had an indoor kitchen, complete with a range and refrigerator, essential for preserving vaccines and food.

When I arrived in Thomonde two months later, the concrete frame was already up. Delva was beaming with pride as we toured the grounds.

"Welcome to Kay Medishare, Dr. Fournier!" He had brought up "masters" for masonry, electricity, roofing, and plumbing from Port-au-Prince. The workers were all Thomondois. They cheered as Delva introduced me in Creole. Ever the politician, Delva made sure they knew that he and I made their jobs possible.

Delva didn't quite make his self-imposed deadline of six months. The windows were not in, the walls needed to be painted, and the floor tiles needed to be set. Still, the showers and the bathrooms worked. The roof didn't leak, and the foliage had grown enough to cool and shade the entire property. We used the tournelle for a health fair with patients lined up on benches, slipping behind sheets hung from ropes for privacy on the stage platform, and then passing by the windows on the front of the house to pick up their medicines. We had over 30 medical students on that trip, plus several Peace Corps volunteers who helped the students with translations, registrations, and organizing the crowds. The students slept in sleeping bags under the tournelle, while the faculty got the bedrooms. Michel, who accompanied me on that trip, shook his head in amazement.

"I've got to hand it to you, Art. I was quite skeptical, but you've really pulled it off!"

"Give Delva the credit, Michel. It was his idea. More importantly, he turned his dream into reality and I have a suspicion he's not done yet. Any nostalgia for the latrine?"

Orphans

With perseverance, I eventually found Pere Luc, who had built his orphanage at Delmas 48. In our first odyssey to find him, we had been off by only six blocks. In Miami, under the vow of obedience to his order and the dominating personality of Brother Paul, he was distinguished from the other brothers only by his accent and dark skin. Back in his native Haiti, he had come into his own. He had started by taking in orphan boys he found in Cité Soleil. When their numbers grew beyond what he could care for in his home, with the help of the Archdiocese of Port-au-Prince and a benefactor from Miami, he founded his orphanage, Bercail Bon Berger (Sheepfold of the Good Shepherd). Soon, with the help of an Irish monk, Brother Charles, he was father and mother to 200 boys.

Stories like Pere Luc's are commonplace in Haiti. With decades of political violence and inadequate resources for health care, orphans are a "growth industry." No one seems to know how big the problem really is. The government has no reliable statistics. The Catholic Church estimates that 200,000 children are institutionalized orphans, and it certainly seems that orphans and orphanages are everywhere in Port-au-Prince. Most are run by religious groups or individuals of goodwill, both Haitian and missionary. With a huge demand and limited resources, the orphanages operate on the margin—taking the maximum number of children they can and hoping that Providence will provide food and clothing for them from month to month. Health care is a luxury few can afford. Some have volunteer doctors who will make sick calls. Most do not. Thus, Project Medishare and its volunteers soon found a welcomed niche in providing screening and preventive services to these children. There are now several orphanages that we visit regularly. Some are newly constructed specifically as orphanages. Others are little more than formerly abandoned buildings. So far, we've found only two with small rooms sleeping 10 children or fewer. Most have large dorms with hundreds of bunks, frequently three tiers high. Dorms like these are designed to allow the orphanages, in their compassion, to care for the maximum number of children they can. Unfortunately, the design fosters the spread of tuberculosis. All it takes is one unidentified active case, and soon most of the children are infected. Coming from desperate poverty, with tuberculosis and HIV frequently claiming the lives of their parents, and with immune systems weakened by malnutrition, new infected children arrive weekly. It's impossible for most of the orphanages to identify and quarantine these sick children.

And yet a visit to a Haitian orphanage is not a trip back in time to some Dickensian Bleak House. Love matters, and at least in the orphanages Project Medishare has worked with, love abounds. I am forever amazed at how outgoing, friendly, happy, and grateful these children are. Even when scars are deep, they somehow heal. For example, Enoch lost his parents to the sinking of the ferry boat *Neptune*. He has not spoken a word since. To his fellow orphans this does not matter. They accept his silence as they do Michael's blindness and Donnell's frequent infections—brothers all. What Enoch can't or won't express with his words is surpassed by his smile, gestures, and touch. The chaos of the playground melds into discipline at a single word from Père Luc or, for that matter, from me. Ron was right: They do teach politeness in the schools, even the orphanage schools.

Delmas is one of Port-au-Prince's busiest neighborhoods, and Delmas 48 is a congested thoroughfare linking Delmas to Museau. Bercail Bon Berger rises up the side of a hill behind a high wall and a large red iron gate. Outside the wall, the sidewalks are filled with market women selling their produce directly from the street, exquisite tomatoes and eggplants, oranges and onions stacked in pyramids. The din of bartering is compounded by the honking of cars and the roar and crunch of large trucks downshifting to ascend the route. Miraculously, these sounds reflect off the wall but do not penetrate the orphanage itself, where the only sound is that of 200 children talking, laughing, shouting, and reciting their lessons. If there's hope for Haiti, it's to be found in places such as this. The courtyard behind the wall is flanked on one side by a newly built school and on the other side by the clinic and chapel. Going up the hill is a small garden and pigpen, a large playground, and a library. The path traverses back and forth, leading to the dorms, the kitchen, the bread bakery, and the rectory for Père Luc and Frère Charles. Play, work and study are not confined to their respective areas, but happen randomly throughout the complex.

I was visiting Bercail Bon Berger with a team of medical students and my younger daughter, Suzanne, to screen for tuberculosis and to hand out toothbrushes. The toothbrushes were Suzanne's idea. She had decided on a career in dentistry in second grade, after a busy but nonlucrative internship extracting her classmates' baby teeth. Intrigued with my stories of Haiti, and filled with her interest in dentistry and the familial volunteer spirit, she conjured up the idea of Project Brossé-Dent (which could be translated as either "Project Toothbrush" or "Project Brush Your Teeth"). Project Brossé-Dent— distributing toothbrushes and toothpaste donated from dentists' offices and giving classes in Creole on dental hygiene—succeeded beyond our wildest dreams. Can you imagine the power of a tooth-

Suzanne and some of her tooth-brushes.

brush to a child who has never had one? Whatever tears were provoked by the prick of the tuberculin test evaporated as each child received his prize, along with a simple message—"*Pou prevni maladi dan pike, bwose dan-w chak jou*" ("To prevent cavities, brush your teeth every day"). In secret I reflected on the irony. My own daughter following in Régis's footsteps. She was 17 at the time.

We stayed for mass after Suzanne distributed the toothbrushes. Père Luc conducted the service in French, with the children singing a capella in Creole, accompanied only by homemade tambours. *Perhaps the French and African heritages of Haiti could live in harmony after all, I thought.* Suzanne later played her flute, accompanied by the boys with the tambours.

As we were preparing to leave, chatting with Père Luc and Frère Charles, four figures caught my eye. Even while they were still in my peripheral vision, their significance registered instantly. My eyes flashed to them and then to Suzanne. She also knew.

Two nuns, dressed in their traditional Sisters of Charity habits, flanked a gaunt-appearing woman carrying a toddler. The toddler had a bushy head of hair that made his head seem twice the size of his body. He was wearing a hand-me-down T-shirt and no pants. Suzanne returned my glance with her eyes and mouth wide open. It was transparent. The mother was dying of AIDS and was giving her child up before she died, at least knowing he would be well cared for. Frère Charles muttered with his incongruous Irish brogue, "Oh, no. Another one. The good sisters bring us two or three a week." Père Luc, Frère Charles, and the two sisters exchanged greetings and then sat down with the mother to do the paperwork. The mother was too weak to stand without the nuns' support. Somehow, the child ended up in Suzanne's arms. The child sensed the tragedy of his separation intuitively and began to wail. Upon affixing her mark to the papers, his mother kissed him and left, aided by the nuns. "Probably hours, at the most days, left," I assessed.

Luc, Charles, and I launched into a long discussion about how the orphanage could cope with what seemed an inexhaustible demand for its services. Meanwhile, the child continued to cry with its head buried in Suzanne's shoulder. Suddenly I realized we were running late. We were past due at another orphanage.

"Okay, team. Pack it up. Say goodbye. We've got to go!" I barked.

"What do I do with him, Dad? Can we take him home with us?"

"No, Suzanne. He'll need to stay here. Give him to me."

I took the child and gave him to one of the older boys. "Take him to the kitchen for some food and then to the playground," I explained in Creole.

Later that evening at the hotel, Suzanne was somber. "What will happen to him?" she implored. "I've still got his tears on my T-shirt."

"Assuming he doesn't also have HIV, and there's a fair chance that he does, although he looked healthy enough, he'll grow up at the orphanage. His memory of his mother will fade, but hopefully it will always be there. Père Luc will be his father, and he'll have 200 brothers."

"It's so sad."

"Every child at the orphanage has a story like that. Multiply that by thousands of orphanages, and that's a small part of the tragedy of Haiti today. Multiply that by the hundreds of countries around the world in the same boat as Haiti and, guess what? You've experienced firsthand something that's probably the most important historical event of your lifetime. Listen. Let's make a vow. We'll keep coming back to check on him. We'll watch him grow. Next trip, you can bring him a toothbrush. Someday when he's old enough to understand, we'll tell him of his mom and her last courageous gift to him. You'll need to work on your Creole."

Protégé

BACK IN MIAMI.

"Pick up, Dr. F. Minority Affairs," chirped Junia.

"Hi, Astrid. How can I help you?" Astrid is associate dean for minority affairs.

"Art, Joe's in academic difficulty. They want to kick him out of school. You've got to help him."

One of the things I liked best about Medishare was how the small number of Haitian-American students at our school rallied around it. Although I had a close relationship with all the Haitian-American students, it was common knowledge that Joseph was a favorite. So when Joseph's troubles surfaced, Astrid knew to call my office. Not just me, but also Michel and André, as well as Junia, were all there for moral support. Joseph came to see us more often than the other students. I think we felt like home away from home for him. Sometimes, he wouldn't even visit for long, just inquire as to news from Haiti from the others or drop in for a minute and ask about our project and how he could help.

According to Astrid, the problem was simply that Joseph had failed a course and then failed the makeup exam—grounds to repeat a year but not grounds for expulsion. But I knew there must be more to the story. When I heard Joseph's voice outside my office talking to Junia, I buzzed her and asked her to send him in.

"What's this all about, Joseph, and how can I help?'

"Well, I had trouble with microbiology and I had to take a makeup. The problem was, I had to take Part I of the boards before the makeup, but they didn't want me to. I had already paid my money, and I couldn't start the third year if I did not pass the boards. So the good news is I passed the boards, including the micro section, but the bad news is they say I flunked the makeup. So that, plus some troubles I had in the first year, means I have to leave school."

"Joseph, do you want me to intervene?"

"No, Dr. F. In fact, you could be part of the problem. They want to know how a 'marginal' student could spend so much time volunteering in Haiti."

"Joseph, don't let anyone ever tell you that you are 'marginal.' You will be a great doctor. The problems are that English is not your first language and you're working three jobs to pay your way through school. But I don't see what the issue is. You passed your boards. That ought to count as the makeup!"

I shared all of this with Michel. "It's like the school has put a zombie curse on poor Joseph," I confided.

"It's a triple curse," Michel came back. "Three strikes and you're out: He's poor, he's black, and he's Haitian."

"They're trying to make an example of him, I know it. The whole medical education department is new, and they're obsessed with grades and board scores. They probably didn't expect him to pass his boards, and that's why they didn't want him to take the foolish test!"

Mark, the new senior associate dean for medical education, was one of my former residents and one of the original collaborators on our AIDS study. Perhaps I could reason with him.

"Hi, Mark. This is Art. I'm calling about Joseph."

"He's in big trouble, Art. Don't try to defend him. He should never have taken the boards without my permission."

"But he passed them! Doesn't that count for something?"

"He's got a big-time attitude problem—always smiling like he knows something I don't know. Plus, his accent is so thick you can cut it with a knife. How is he going to present his patients in his third year? They'll eat him alive on his clinical rotations."

"Mark, he got into academic difficulty because he's working three jobs and didn't even start learning English until he was 14. Can't we cross the accent issue when we come to it?" I was inwardly boiling, trying to restrain myself.

"Hey, it's my job to uphold the school's standards. This guy just doesn't make the grade."

"I'm going to advise him to appeal."

"That's his right."

Every student threatened with academic suspension has the right to appeal to the dean. This was Joseph's last hope. I counseled him not to despair. The dean was a straight shooter. He was fair. When he heard the whole story, things would be set straight. Astrid and I would ask to be present.

Appeal day arrived. It was strained sitting around the table in the dean's conference room with Astrid, Mark, Joseph, and the associate dean for students, waiting for the dean to join us. Astrid and I had counseled Joseph to be himself and not be nervous, but the small beads of perspiration on his forehead and the absence of his usually irrepressible smile were foreboding.

The dean entered from his office, almost like a judge entering a courtroom. After he seated himself, he looked at Astrid and me and told us we were welcome to attend the proceedings as a show of support for the student in question, but we were not allowed to speak and were not allowed to ask questions.

"Uh-oh," I thought.

Mark presented the recommendation of the Promotions Committee that Joseph be dismissed, citing academic difficulty and insubordination for taking the boards when not authorized to do so.

The dean asked Joseph if there was anything he could say that would demonstrate that he had the capacity to meet the school's academic standards.

Just say the magic five words—Yes. I passed the boards, I thought as I stifled myself from an outburst. Joseph tried to get something out but could only stutter and stammer with his Creole accent. The dean asked Joseph if he had a plan to return to academic good standing. Again, all Joseph could do was stammer incoherently.

Silence suspended the proceedings for what seemed an eternity as the dean wrote some notes and then made his pronouncement. Joseph was expelled from our school. To be reinstated, he had to make up and pass the course in question. That meant he'd lose at least a year. He also had to demonstrate his capacity to acquire biomedical knowledge by obtaining an advanced degree in one of the biomedical sciences. Finally, he had to enroll in an accent reduction class.

The last Herculean task pushed me over the edge. Clearly, Mark had met with and prejudiced the dean prior to the appeal and had baited Joseph to bring out his stammering. This was worse than a zombie curse. At least if you are a zombie there's a small hope for resurrection and escape. This was an execution. No one could possibly meet the requirements for reinstatement. The dean instructed Joseph to meet with Mark and the dean of students immediately after the meeting to review the terms of his expulsion and possible reinstatement. I passed Joseph a note telling him to come by my office after that meeting. Joseph came by later that afternoon, slumped in the chair on the other side of my desk, and buried his head in his hands and arms.

"What are we going to do?" I asked.

"You can't do anything, Dr. Fournier. I have to bend to their wishes. I'll enroll in a master's in public health course."

"The thing that bothers me the most is that stupid accent reduction requirement. That's really off the wall and uncalled for. That didn't come from the dean."

"Its okay, Dr. Fournier. I want to get back in school. I want to get my degree, complete my residency and go back to Haiti and make a difference."

"I have a friend of my family who's a speech pathologist. I can ask her if she'd take your case."

"Thanks. That would be helpful. But I don't think I can come around anymore. They don't like it that I come to you, and they don't like it that I've gone so often to Haiti."

To me Joseph's expulsion from the school was just one example of how the curses of stereotype, blame, and conventional thinking were affecting not just our patients but the profession of medicine itself. In Joseph's case he had been stereotyped as a poor student, blamed for taking the boards without permission, and ostracized without a fair hearing. In my opinion the school had cursed not only Joseph but also itself in the process—enslaving itself by an obsession with grades and authority. In reality, what Joseph needed was a little remediation and a scholarship! Even the dean had become ensnared in the trap of going through the motions of doing what he had to.

When I was a medical student, my school was "pass-fail." A pass-fail system discourages competitiveness among students and encourages learning for the sake of learning, not for the sake of grades. When I first came to the University of Miami, it also was a pass-fail system. However, whether because of insecurity or a desire to procure its graduates the "best possible residencies," the University of Miami School of Medicine had evolved through the years from pass-fail, through letter grades, to now a numerical grading system. Prior to going to Haiti, I had protested the grading system as the antithesis of learning.

After returning from Haiti I became even more vocal about our grading system. "We ought to be evaluating our students on the knowledge, skills, and aptitudes needed to be good doctors and give them extra credit if they volunteer for projects like our health fairs or Medishare," I argued to the school administration. "Not on their test-taking ability or their capacity for rote memorization." But the school felt compelled to grade our students to the *nth* degree, believing that it was necessary if the school was going to compete with the best schools in the nation for the best students.

The students could not help but get caught up in this obsession. For the majority, grades became the be-all and end-all of their existence. The school fueled the flames of this competitiveness by actually posting the grades of every test for all to see. Meanwhile, special students like Joseph fell through the cracks. If they failed, we blamed them for their failure, rather than asking ourselves what's wrong with our system. Even those who succeeded in the system paid a price—medical education had become four years of a numbing, competitive, cutthroat grind, to be followed by three to five years of further "training." So many of our students entered the system with idealism, and so many left with cynicism and burnout—medical zombies!

My school's obsession with grades was probably not that different from most medical schools in the nation. In fact, the entire profession of medicine in the United States seemed to be undergoing a process of zombification.

Over the past 20 years, Medicine had shifted from a cottage industry that physicians controlled to a corporate entity in which doctors were little more than glorified assembly line workers. Medicine was no longer a profession but a business. Health care was no longer a right; it was a commodity. The precious doctor-patient relationship had been torn apart by managed care and stood on its head by malpractice issues. In response, more and more physicians were withdrawing, becoming defensive. The patients we were supposed to be serving became the enemy. Increasingly, doctors were dropping out. Older doctors were retiring early and recent graduates were choosing careers like radiology or pathology, with little patient contact, or dermatology, with little responsibility for nights and weekends and therefore few conflicts with life style. Hardly anyone was choosing a career in primary care. And in both cases, generalists and specialists, everyone seemed to be just going through the motions.

Making the problem worse, as I mentioned earlier, was modern medicine's enslavement by technology. Cardiologists can no longer commit to a diagnosis of a heart murmur without an echocardiogram. Neurologists can't diagnose a stroke without a CAT scan. The old art of medicine—talking to patients, figuring out what's wrong, and fixing it—had died. In the process, actually *caring* for patients had also died. The replacement? Robotic surgery, high-tech scanners, productivity quotas for doctors, and precious little time for patients. Health care in America was becoming "The Night of the Living Dead."

Medishare had originally gone to Haiti to help the Haitian people. My first trip awoke me from my own zombie curse. It was becoming clearer and clearer to me that perhaps Haiti was helping me more than I was helping it. Perhaps experiences in places like Haiti could help some in my profession shake off the curse cast upon our profession. I had witnessed firsthand how touched our students were by the heartfelt "*Mesi, ti dokte*" they received from every patient after every service they performed. Medicine didn't have to be robotic, "going through the motions," or uncaring. There was more to being a medical student than passing exams and more to being a doctor than making money. If only our school of medicine would see it that way.

The story of Joseph does, however, end on a happy note. It took him three years, but perhaps to the surprise (and consternation) of the school's administration, he completed all of the dean's prerequisites for reinstatement. He graduated three years ago. I cheered wildly for him when the dean handed him his diploma. He is now completing his residency in the family medicine training program here at the University of Miami. It was Joseph's Haitianess—the dogged determination to climb one mountain after another until he reached his goal—that saved his career. If Sisyphus had been Haitian, he would have gotten that rock over the hill.

Benediction

Spring 2001. Driving on Haitian roads demands the complete attention of all of one's senses. On the one hand, the questions of my students were somewhat of a distraction. Potholes, pedestrians, and *polis estasyonnen* (speed bumps) lurked ahead, while in the opposite direction tap-taps of every color and dimension rushed toward us bumping, jerking, and narrowly missing us at every turn. On the other hand, I'd never felt more alive. Nothing seemed impossible.

We were driving to Croix de Bouquets to read tuberculin tests at an orphanage. In seven years, Medishare had grown from a small volunteer organization to a charity sponsoring major programs in community health in Thomonde and in training Haitian doctors to be family physicians in Cap Haitien and Pignon. At the same time, we maintained an ongoing commitment to provide health care services to a cluster of orphanages in the vicinity of Port-au-Prince. Support for our charity had grown steadily. Initially funded predominantly by contributions from Barth and to a lesser extent by me, we were becoming more and more successful with our art auctions and with grants and donations from the community. Furthermore, we impressed on our students the need to raise funds—at least what was required to pay their own way.

A cohesive plan was starting to evolve. The first class of Haitian residents was at work at Cap Haitien and Pignon, and over 100 medical students had volunteered for Project Medishare. Between the orphanages, the residency program, and Thomonde, we were poised to make a difference. There was no denying that Haitians were still dying in boats trying to get to Florida, that AIDS was still creating orphans, that the news out of Port-au-Prince was perpetually bad, and that the "powers that be" at our medical school still didn't get it. We were making a difference. The finally completed guesthouse in Thomonde exceeded everyone's expectations, not just in terms of comfort but also in expanding our capacity to work in the village of Thomonde and its surrounding communities. Students from other medical schools, coming to Miami for fourth-year experiences called externships from as far away as Saskatchewan, heard about our Haiti experiences and signed on.

"What about the elections?" came a voice from the back of the van.

"They're going to happen," I answered.

The Miami press was all wrapped up in the impending legislative and municipal elections. Was Aristide stalling, so that one combined presidential and legislative election would sweep Lavalas into power at all levels?

"None of our business," I told the students. "Too much of Haitian politics starts and ends in Washington. What this election flap is all about is whether there will be a 'globalization' of the Haitian economy. Aristide was elected by promising jobs for the poor and taxes for the rich. After the coup and his expulsion, accommodation with globalization was the price he had to promise in exchange for a United Nations intervention to return him to power. The Haitian peasants couldn't care less about the World Bank or globalization of their economy. Furthermore, the bloom was off the rose of globalization. It was not working the miracles it had promised, and its cost is being carried on the backs of the people."

Meanwhile, as Paul had explained, 50 years of American foreign policy toward Haiti had been shaped by the fear that Haiti would go communist. Cuba had gone communist, and according to theory, Haiti, even poorer, was certain to follow. But anyone with a modicum of knowledge of Haitian history knew Haiti would never become a communist country. To Haitian peasants the land was their birthright, the gift of their ancestors' revolution. They'd never give it up. So for that misreading of history we supported Duvalier and the generals. My advice is, "If you want to help Haiti, ignore the politics and focus one-on-one on our humanitarian mission. Given enough time without outside influence, the Haitians will fix their own politics. It won't be easy, but it's the only way."

"I don't understand this birthright thing, Dr. Fournier," came a challenging voice from the back of the truck. "And this liberation theology preached by President Aristide. It sure does sound a lot like socialism."

"In 1815, Haiti's leaders tried to revive the plantation economy. They enacted 'land reforms' that would have tied the former slaves back to the sugar fields and a state of serfdom. The peasants voted with their feet. They took off into the mountains, carved up the land among themselves, and passed it on from generation to generation. In effect, they traded a subsistence existence for their freedom. Aristide and his liberation theology have to be seen in the context of 200 years of struggle between the poor and the elite for what they both consider their birthright. But trust me—the Haitian peasants will never give up their land. The problem is, with the average family having five kids, the birthright gets smaller and smaller and subsistence gets harder and harder."

Silence descended on the van as the students pondered my diatribe. It was great to see how much they learned in only a week, not just about medicine but about Haiti, the world, and themselves. Past the wattle and daub huts, the piles of watermelons, the hand woven baskets, and the continuous stream of life on the Route Nationale, we raced from the north toward Croix des Bouquets. As we passed the remains of Ibo Beach, I pointed out where tourists had most likely introduced HIV to Haiti, then the salt flats where the bodies of Duvalier's victims were left for the crabs to scavenge, and the trash dumps where children still forage for scraps to this day. "We can make a difference," I reminded myself.

"Dr. Fournier," called another unseen voice. "What brought you to Haiti?"

"That's a long story and it's only 15 minutes to Croix des Bouquets. Let's save that one for dinner tonight. Have I told you the secret of the zombie curse?" Fifteen minutes was not long enough to capture their imagination with tales of death and resurrection, freedom and slavery. *Thank you, Haiti*, I thought to myself. *I had the zombie curse of a comfortable life in the United States. For all my good intentions, I was in a coma. Previlus, Belony, Theophile, Marie—I'm trying. Régis—I've not forgotten. Hopefully, we will make a difference.*

I proved to be naive in my attitude toward Haitian politics. The elections were contested, and that led in certain legislative districts to a political impasse that paralyzed the government and froze international aid to Haiti. Conditions around the country deteriorated, and people once again started heading out in boats, except this time few made it to Miami. Most were either intercepted by the Coast Guard and returned to Haiti or were lost at sea. Haiti became a political football again, and the people, not the government, felt the pain.

To my shock, Delva lost his bid for reelection as magistrate of Thomonde. Delva's losing that election was about as improbable as the Red Sox losing an election to choose Boston's favorite baseball team. To make matters worse, the winner, a Lavalas insider, lived in Port-au-Prince. He didn't even live in Thomonde!

"It's all right, Dr. Fournier," Delva said, after I expressed my condolences. "He's my cousin! Besides, now I can devote all my effort to the health of Thomonde."

Delva wasn't trying to create a consolation prize for himself. The Green Family Foundation, a Miami benefactor, had funded

Medishare's proposal for a community health project in Thomonde. In partnership with Paul, we planned to renovate the dispensary and hire Haitian doctors and nurses to bring full-time health care to Thomonde. We also planned to hire a small army of community health workers. Paul would train them in outreach and direct observed therapy, a simple but effective way to make sure sick patients get their medicines. It turned out that in his youth Delva had studied public health and had trained as a tuberculosis control worker. So he would lead our team. It was just like Delva—take lemons and turn them into lemonade.

For several months a line had been drawn in the sand at the medical school over Project Medishare. The provost, the general counsel, and the senior associate dean for medical education were on our case over the possible liability accrued by students volunteering in Haiti and their perception of the danger our students might face while working in Haiti. These concerns were fueled by a continuing string of sensationalist articles about the political instability in Haiti in the *Miami Herald*. It seemed that negative articles always appeared in our local newspaper one or two weeks before each trip's departure. On the other hand, we had Barth's clout behind us. He was probably the most successful fundraiser the University of Miami had ever seen. In addition, we had the example set by Paul and his relationship to Harvard and our students, who were doing important work on tuberculosis among orphans, sickle-cell anemia, and women's health. The dean was noncommittal for several months. Who could blame him, with the challenges facing the medical center? The grant from the Green Family Foundation tipped the balance in our favor— $250,000 per year for three years to improve the health of Thomonde and a parallel grant to the department of pediatrics to start an international pediatric infectious diseases program. A condition of the award was that the issues between Medishare and the school had to be resolved.

The grant did more than give Medishare leverage at the school. It gave Medishare a permanent health presence in Thomonde, allowing us to break through the limits of what we could accomplish with only episodic visits. The key was not so much the renovated clinic or the Haitian doctors or nurses who would be there full time (although these were all critical advances) but the 36 community health workers, handpicked by Delva.

He chose them from every corner of the commune—men, women, young and old, former farmers, market women, teachers, and students. The grant provided them with mules and motorcycles that enabled them to visit every household of every district— Thomonde, Boucan Carrè, Savanette, Tierre Muscadet, and Baille Touribe. Trained to a remarkable degree of medical sophistication by Paul, they could identify the ill, bring them to the dispensary for diagnosis and treatment, educate them as peers, and ensure that they would take their medicines. It was health care by the people, of the people, for the people. We paid each worker about $100 per month, about five times the amount they could earn by farming. With Delva leading them and Paul inspiring them, they took their work with utmost seriousness.

Delva arranged for the Catholic bishop to come down from Hinche to bless our guest house and clinic. Actually, the guest house today is no longer just our guest house; it serves as the health care headquarters for our community health workers. It's now the most substantial home in Thomonde, surpassing that of the funeral director. Everything in Haiti has meaning, and the symbolism of this proprietal comeuppance was not lost on the Thomondois or, for that matter, on me. Death is a growth industry in rural Haiti, and in practically every town the largest and most prosperous-looking home always belongs to the funeral director. This is no longer the case in Thomonde. Health now has the upper hand.

The house blessing ceremony included a visit by Paul and his staff from Cange. The ceremony was scheduled for 4:00 in the afternoon, and we spent the day prior to the inauguration giving health care to the citizens of Thomonde. Barth came—his third trip to Haiti with Medishare—and differentiated from neurosurgeon to primary care physician. After conceptualizing Project Medishare and financing and participating in the first trip, he had pretty much turned the project over to me. This was certainly understandable. In addition to running the neurosurgery department, he had launched a campaign to find a cure for paralysis. Now, however, seven years into the project, perhaps because his kids were now volunteering in Medishare, he was once again engaged. We had 30 medical students with us. They had worked hard all day, but filled in the back rows of the tournelle during the benediction, surreptitiously sipping beer and marveling at the event.

Paul arrived early and really hit it off with Barth and the students. He and Barth talked about building a day hospital on the property next to the guest house. The bishop said a high mass in French. Afterwards, Barth, Paul, and I were asked to speak. Paul worked the crowd with a rousing speech in Creole, celebrating with the Thomondois their participation in living a healthy life.

After mass the bishop passed through the house and blessed it with holy water. The Thomonde brass band, usually relegated only to funerals, played several celebratory songs, including the Haitian and American national anthems. Medical students and the citizens of Thomonde intermingled, drinking beer and home-brewed clairin. The generator hummed, providing light and music well into the evening. Even Paul let his guard down, socializing with the students and sipping enough beer to start telling jokes and pose for pictures. The Peace Corps volunteers fueled the celebration with a homemade rum punch.

I slipped away to a quiet corner of our property to play the "rock game" with some of Thomonde's children. I always play the rock game with the children. They count on this and sit patiently, 20 to 30 at a time, every evening I spend in Thomonde.

The game is simple. First, I search the ground for a small, shiny rock. Then I line up the children in a row, sitting on their haunches. I place the rock in a hand behind my back and then offer them my two closed fists to choose which one contains the hidden rock. If they choose the hand with the rock, they win. If they choose the empty hand, they lose. I go down the line, and each child gets to choose. Each child gets lots of advice from the others when it is their turn, and each choice, right or wrong, is greeted with peals of laughter. *"Ou genyen"* ("You win"), *"Ou genyen. . . . Ou pèdi. . . ."* ("You lose"), down the line I go. I pass through the line several times, so that each child wins at least once, and then turn the game over to them. That night the rock game went for hours. Haiti, land of contrasts. For all the misery that surrounds them, these children found joy in guessing which hand holds a small rock. Not just a little joy, mind you, but unabashed, unrestrained joy—more joy than could possibly be produced by a million video games. As much as I keep trying to give to Haiti, Haiti—particularly its children—keeps giving even more to me. When I awoke the next morning the children were there again, ready to begin at first light. I passed the rock to a young girl who seemed precocious, and I took her place in line, pretending to be a kid again, but this time a Haitian kid. I won each time.

Later that morning I was standing under the tournelle, sipping a cup of coffee. It seemed strange that I was the only one up at 8:00. Granted, I had turned in early while the celebrations were still going on, but everyone knew we had to be on the road by 8:30, as patients would be waiting for us at Savanette and Tierre Muscadet. The students were huddled in their sleeping bags under the tournelle and Barth was locked inside his guest room.

Finally, Gina, one of the student leaders, roused herself and came over to me, following the scent of freshly brewed coffee. "What happened to everyone?" I asked, innocently.

"Um, Barth thought it would be a good idea to take us to a Voodoo ceremony. He asked Delva to arrange it."

I cringed. *Barth, what were you thinking?* I thought to myself.

"It didn't go well," Gina continued. "It was pretty staged, very touristy, mostly dancing, singing, and drum beating. A lot of candles. And then the Voodoo priest wanted to be paid, and when Barth didn't pay him right away, it got a little ugly for a while."

"I'm almost afraid to ask what happened next."

"Oh, Delva smoothed things over. He probably paid the priest a small amount of cash. In any event, Delva gathered up some musicians from the brass band and we danced until 2:00 a.m."

Gina left me to rouse the other students. About half an hour later, Barth emerged from his room. "You missed a good time last night, Art," he said sheepishly.

"Barth, I heard what you did. That was awful. Cultural voyeurism. What made you think you could buy a Voodoo ceremony?"

"Well, I paid for my kids' bar mitzvahs, didn't I?"

"That was different. You were lucky you had Delva to bail you out. It could have given Medishare a black eye with the people!"

"Boy, you're getting as bad as Paul."

Discovery

Spring 2002. MY TEAM AND I WERE Debriefing after a busy day of health fairs at orphanages. Since our first visit, the orphan problem had captured my heart. Most orphanages were started by people of good will, but their medical sophistication varied, as did their resources. Some had Haitian doctors on call for acute illnesses; some had a nurse on duty. In general, however, health was prioritized well below food, shelter, and education. Particularly poignant were those orphanages that took in handicapped and abandoned children.

Haitian family structure is quite strong. It takes a lot to drive a Haitian family to abandon a child, usually a combination of desperate poverty, severe disability on the part of the child, and impending family collapse. Not all orphanages accepted abandoned or handicapped children. That morning we happened to visit two that did.

Debriefings were an integral part of Medishare's method. Every evening we reviewed the events of the day, presented the unusual problems we'd seen, identified what we did right and what we could do better, and addressed any social or cultural issues that might have surfaced.

It takes dedication to examine a handicapped orphan in Haiti. They seemed so hopeless; it was hard not to get depressed. The orphanages had sequestered these children in separate rooms away from the healthy children. Most were wasted, as they could not feed themselves and required forced feedings, which were difficult for their *matant* (literally "my aunt"), the Haitian women providing care, to perform. Their limbs were spastic, their comprehension was minimal and, regardless of their physical age, their functional level was a year or less. I attributed their condition to birth complications–lack of oxygen, maternal hemorrhage, or prolonged labor.

I had a small but enthusiastic team—Russ, Parul, and Rob, all first-year students, plus my sister-in-law, Nancy, a pediatrician from Cape Cod. The discussion focused on the visual problems of four of the children.

"Dr. F., four of these children have cataracts. Why is that?"

"Great question, Rob. Usually, we see cataracts only in old folks." Actually, I was dumbfounded by this finding. "Perhaps it's a tropical medicine thing. We'll need to research it when we get back."

"Did they [the children with cataracts] have any other findings?" asked Nancy.

"Yeah," volunteered Russ. "There was not one of them that was normal. Kind of funny-looking."

A long pause ensued.

"Congenital rubella. Perhaps they have congenital rubella," Nancy speculated.

My sister-in-law had provided us with a brilliant insight. Rubella (German measles) is a benign viral infection that causes a few days of mild fever and rash, unless one happens to be infected early in pregnancy. In that case the virus can cause cataracts, deafness, retardation, abnormally small heads, and heart defects in the fetus.

Nancy's insight awoke me from the fatigue of a day in which we saw 300 children. This was really important. We'd missed it on the first pass because no one in the United States ever sees congenital rubella any more. It's easily preventable with immunization. Haiti, however, was the only country in the hemisphere that did not immunize against rubella.

It was a moment of awakening similar to my first visit to Cité Soleil. Far beyond the four children with cataracts, rubella explained probably a hundred or more handicapped children I had seen in Haiti. Even that young woman in Zetoit probably was a rubella victim. And it all could be prevented by a simple, cheap vaccine. All of this misery and suffering didn't have to be. These children could have lived normal lives. Instead, they'd been sentenced to a life worse than that of a zombie, simply by the accident of their birth.

"I think Dr. Nancy is right," I chimed in. "You all have made an important discovery. When you get back, you should write it up. I'm impressed. There aren't that many first-year medical students in the entire United States who can use an ophthalmoscope well enough to see cataracts in adults, let alone kids. Let's pass by tomorrow and see if these kids meet the other criteria for congenital rubella. It's a clinical diagnosis; there are no lab tests. We'll need to measure their head circumference and test their hearing." The discussion then turned to tuberculosis, malnutrition, and worms. Before breaking up, I asked the students if they had any questions about Haitian history or culture.

"Tell us about Voodoo, Dr. Fournier," whispered Parul, looking furtively from side to side to see if the waitresses were listening. "I noticed some children today with amulets around their necks."

I paused and contemplated how heavily I wanted to delve into this question. "Think of Voodoo and Christianity as the yin and yang of Haitian spirituality. If the Christian God created heaven and earth and keeps the stars twinkling in the sky, the Voodoo spirits, the African gods, are there for your personal intervention in matters of love, health, and family. If Christianity is Apollonian, Voodoo is Dionysian. The church has its liturgy and its ritual, Voodoo its ecstasy and possession."

"Possession?" asked Rob through a blank stare.

"Sure. It's almost sexual. The spirit of the *Lwa* actually 'mounts' the supplicant, leaving him or her in a state of exhaustion."

"Wow, that's pretty powerful stuff. Much more exciting than my Sunday mornings at church," replied Russ.

"And then there are the saints," I continued. "A perfect blend of Rome and Africa. Saint Patrick is an incredibly popular Voodoo saint. Any ideas why?"

The group looked at me and each other. Silence.

"*You* look at a picture of Saint Patrick driving the snakes out of Ireland and see the saint. The Haitian looks at the same picture and sees the snakes—Damballah, the spirit of fertility and health and magic, the same symbol we in medicine conceptualize in our caduceus. Then there's the Virgin Mary as Erzelie, the goddess of love, and Baron Samedi, the god of the underworld and so on."

"So what percentage of Haitians are Catholic, what percent Protestant, and what percent practice Voodoo?" inquired Rob.

"It doesn't have to be exclusive. A Haitian sees no conflict between being a practicing Catholic and a practitioner of Voodoo. The Protestants are a little different. They tend to lump Voodoo in with devil worship and black magic. In reality, Voodoo and black magic are separate but related things. Most of my Haitian colleagues say Haiti is 85 percent Catholic, 15 percent Protestant, and 100 percent Voodoo."

The candles were burning down, and the students' heads were starting to nod. They'd been up for 14 hours and we needed to get started at 6:00. "Okay. Enough for tonight. Let's turn in."

War and Peace

THE FOLLOWING EVENING I WAS sitting on the dock at Habitation Labadie, sipping a rum punch and watching the sun set when one of the hotel employees ran up to tell me I had a phone call. "Who is it?" I asked, surprised that someone had found our little hideaway.

"Your dean, I think."

"Uh-oh. Um. . . . Would you mind rounding up the medical students who are with me and ask them to meet me at the phone?"

Actually, it wasn't the dean, but Mark, the senior associate dean for medical education.

"Hi, Mark! How'd you find us?"

"I got your number from Larry at the Center for Haitian Studies. Where are you?"

"Exactly where I said we'd be if there was trouble. We're in the walled and guarded resort I told you about in Labadie."

The simmering issues between Medishare and the medical school were coming to a head. On the one hand, there was Mark, the general counsel, and the provost, terrified by the never-ending series of articles in the *Miami Herald* focusing on Haiti's political violence and concerned about liability and their responsibilities *in loco parentis*. On the other hand, there was Barth, the school's most effective fundraiser, myself, and a growing number of other faculty members and students who came to Haiti, had life experiences, and, more importantly, transformed those experiences into grants, projects, and peer-reviewed articles that made the project valuable to the university.

Our residency training program in Cap Haitien was the first grant the school ever received from the Open Society Institute, the foundation established by George Soros. An article describing the project had been published in *Academic Medicine,* and our paper on screening for tuberculosis in orphanages won a prize for best student paper from the *American Journal of Preventive Medicine.* Unfortunately, liability was a huge issue for our school, and the photos published by the *Herald* just before we left were particularly graphic. One photo of a demonstrator with his hand shot off graced the front page. So that day I got a call from Mark telling me that the provost wanted us to cancel our trip. My protestations that it was too late to cancel, that people were counting on us, and that we weren't going to be in the capital where the violence was centered but rather in the countryside, at the invitation of the people, fell on deaf ears. His opposition to the trip was no small issue, as Mark could make the students' lives miserable and the provost has the power to fire even tenured professors. My final parry—we were volunteers, traveling on our vacations with no ties with the school—produced a compromise. Mark would "allow" us to go, but at the first sign of trouble we had to seek safe haven.

That is how we got to Labadie, the pristine peninsula owned by Royal Caribbean Cruise Line on the other side of the mountain from Cap Haitien, where we had conducted a health fair a few years before. Royal Caribbean is so protective of it that it is walled off and guarded with private security forces. All the people there, including the guards, knew me. There was no safer place on earth. It was our escape valve.

The day started in Leogane at a subsidiary orphanage/farm set up by Père Luc. The students were continuing their physicals on children ranging in ages from 2 to 6.

"Hey, Dr. F!" yelled Russ, "Would you check out this kid's heart? His whole chest wall is buzzing." I listened for all of five seconds and then called the other students to examine six-year-old Dieudonel.

"He's got a hole in his heart—a ventricular septal defect (VSD). What you're feeling is called a thrill. Congratulations, Russ. A good pickup. It probably explains why he falls off the growth curve. We may need to bring him back to Miami to get it fixed."

My dissertation on the differential diagnosis of heart murmurs was interrupted by Delva, who had been waiting in our 4 × 4, listening to the radio. He motioned to me to leave the students and come have a seat next to Père Luc.

"There's trouble on the road back to Port-au-Prince in Carrefour," he whispered in Creole. "A little manifestation, a few burning tires. That's all."

"What should we do, Delva? I promised our school that if there was trouble we'd hide out in Labadie."

"I don't think this will escalate to the level that you should change your plans," said Delva. Père Luc invited us to spend the night at the orphanage. I thanked him but explained that wouldn't solve my problem with the school. The irony was that the only road to the airport, where we could catch a plane to Cap Haitien, passed right through the neighborhood where the protests were occurring.

"Go back to your students and the children," counseled Delva. "I'll keep listening to the radio."

"Dr. F! We've found two more children with cataracts!" exclaimed Rob.

"Make sure you take pictures," I suggested. "You're a lucky group of first-year students. How many medical students in the whole United States have seen, much less diagnosed themselves, a VSD and congenital rubella in the same day? Let's get ready to go. We may have to fly to Cap Haitien."

At 2:00 p.m. Delva reported that the radio claimed all was quiet in Carrefour and Port-au-Prince. In Creole, r's are very soft. So the French *carrefour* ("dock" or "quayside") becomes *kafou*—literally, "crazy car," which is exactly what it is on most days. The most densely populated section of the most densely populated country in the hemisphere flushes its traffic into the one functioning west-east road, creating traffic jams, pedestrian jams, and creative route seekers hoping to beat the traffic into and out of Port-au-Prince. There have been times when traversing this stretch of road from Leogane to Port-au-Prince would take four hours. Today, miraculously, we passed through in less than one hour.

"The demonstrations worked in our favor" was Delva's opinion. Had we come by two hours earlier we would have lost our tires, if not more. We arrived at the general airport and were soon whisked away for the short plane ride to Cap Haitien. Then, after a 45-minute bone-jarring ride around the Cap Haitien mountain and some sweet talking of the guard to let us in, we arrived at the beach at Labadie and awaited the water taxi that would take us to our hotel.

After only four days in Haiti, the students were not prepared for the culture shock of the manicured, tourist-ready Labadie. "Did you see the Caribbean Market building with the sign in English? said Rob, sarcastically. "They should take the tourists over the hill to see the iron market in Cap Haitien. Now that's a market!"

"It certainly is beautiful here," offered Parul. "Who knew?"

"Columbus described this whole coast in the log of his first voyage. The wreck of the Santa Maria is five miles due east of that rocky point over there." I gestured toward Point Honoré. No one believed Columbus's descriptions—they thought he was selling real estate. Actually, his description was quite accurate and—except for the tourist stuff—what you are seeing now is pretty much what he saw over 500 years ago."

"Wow, Dr. Fournier. This morning Father Luc's orphanage. This afternoon Carrefour. This evening Labadie. Haiti, land of contrasts.

They're not going to believe us when we get home," exclaimed Russ. "Thank you."

Russ's words proved to be ironic, although disbelief does not do justice to the hardened opinions back home. The early evening's news had broadcast the morning protest back in Miami, hence Mark's phone call.

"We want you to come back home, Art. We're concerned about your safety."

"Mark, we're up here in this resort. We've got the whole hotel to ourselves, and there isn't another person around for miles. The students are sipping rum punches after a hard day's work. We're due home in two days anyway. If we come home early, we'll have to return to Port-au-Prince. That's where the trouble is! Plus we'll have the added expense of changing our tickets, on top of what I'm already paying extra to stay in this nice hotel. Here—talk to the students."

The students, summoned up from the beach, interrupted, shouting into the receiver.

"You don't understand! It's paradise here!"

"You wouldn't believe the stuff we learned today."

"We feel safer here than in Miami!" "Here. He wants to talk to you again."

"Art. The provost wants you home. The dean wants you home. I want you home. Come home now or we'll never allow another student to go to Haiti ever again."

"Well, I think you are all overreacting, but if you insist, we can fly directly back from Cap Haitien. Is the school willing to help us with the added expense?"

A long pause and then, "Yes. Come home tomorrow."

The students were frustrated and despondent for the rest of the evening. "Just when we were getting good!" one exclaimed.

"What about the other orphanages we were supposed to visit? I don't understand the attitude of our school," another added.

"Things will work out," I counseled. "We're in a war for the health of the people in Haiti. This is just a small tactical retreat. Live to fight another day. Have another rum punch. We'll be back with another team in three months."

Medishare did, indeed, return to Haiti three months later, despite continuing political turmoil and perpetual bad press. That trip and the next several trips happened over the objections of the dean and the provost. It threatened to get ugly.

Au Cap

By 2003, ENTERING ITS NINTH YEAR, Project Medishare had evolved from a small group of volunteers to a full-fledged charity. We had been successful enough in our fundraising to be able to hire a full-time executive director to develop year-round programs. Our mission was also evolving. Barth's original concept had been a somewhat limited project: Concerned faculty at the University of Miami would provide technical assistance and training to Haitian doctors and nurses while South Florida health institutions donated surplus usable equipment and supplies. As we became more engaged in Haiti, we morphed into something more unique and significant—in effect, a multidimensional human resource project, developing the potential of our own students and faculty to be compassionate and effective doctors while developing Haitian human resources that would contribute their own solutions to Haiti's health problems. In this regard (I wish I could say we planned it this way, but actually it just kind of happened) we had created two distinct human resource programs. In Thomonde we were working directly with the people to improve the health of the community through peer educators and facilitators—people Paul referred to as *accompagnateurs* (companions), similar to the barefoot doctors of China. At the same time, in

Hôpital Justinien, Cap Haitien, home of the Family Medicine Residency Program.

Cap Haitien and Pignon, we were training Haitian doctors in what was, for Haiti, a revolutionary approach to health care.

A historic and revolutionary program deserves a historically revolutionary setting. Cap Haitien, the principal site of our family medicine residency program (we also used Dr. Guy's hospital in Pignon for training), was the colonial capital of Saint Dominique. In 1791, slaves in the surrounding countryside, taking advantage of the chaos created by the French revolution, rose up against their masters. Legend has it that a Voodoo dance was the sign the slaves chose to signal the start of the uprising. The revolution lasted 13 years before Napoleon withdrew his troops and recognized Haiti's independence. The costly, failed expedition to recapture France's wealthiest colony forced him to sell his Louisiana territories to the United States. Haiti became the first black republic and the second (after the United States) republic of modern times.

The historic old city, once referred to as "the pearl of the Antilles," is now in a sad state of disrepair. Balconies have collapsed, and tile roofs have caved in. Practically every home needs a fresh coat of paint. Peasants fleeing the desperation of the countryside have squatted in the hills and along the river that leads to its port, transforming it into a mini Port-au-Prince, with slums of cardboard and tin shacks swarming with thousands of people. Chaos reigns in the charcoal and iron markets. Its beautiful natural harbor is visibly polluted and its carrefour littered with trash.

Its charm is still apparent, however, to the discerning observer. The hotel I stay at—the Roi Henri Christophe—was built in 1724 and was part of the original governor's palace. When I walk from the hotel to Hôpital Justinien, past the crumbling colonial houses, past the Place D'Armes, where Mackandal, the first slave to attempt a revolt, in 1715, was burned at the stake and where Ogé, in 1792, was broken on the wheel, I'm reminded of the history of Haiti's tragic birth. The cathedral was built in 1630, the year my hometown, Boston, was founded. In 1630, Miami was an Indian village and would remain so for almost 300 more years.

In the mountains on the other side of the Grande Plaine du Nord, Christophe's fortress, the Citadelle, juts above the geologic skyline. Not far is Breda, the plantation where Toussaint L'Ouverture, Haiti's first leader, was born and Vertierres, the site of the last rearguard action of the French before they abandoned their former colony. Their place in Haiti's history is not lost on the Capois, who fight their own rearguard action against the continuing neglect and decay of their city. André was right to choose Cap Haitien for our family medicine training program. Hôpital Justinien is arguably the best-functioning government hospital in the country. Port-au-Prince is a snake pit of politics and intrigue, even in the health sector. *Au Cap*, André was on good terms with the hospital director and the regional health minister. They agreed to pay for the program's staff and the residents' stipends. André had trained as a surgeon at the university hospital in Port-au-Prince. During his year of social service in the town of Fort Liberté, he realized how poorly his training, although technically quite competent, had prepared him for the vast majority of problems that beset his patients. He used to go up on the old fortress wall that jutted into the harbor and contemplate what he could do for his beloved Haiti. The answer he came up with was to immigrate to the United States, become a family practitioner, and then return to Haiti and establish the discipline there.

André was therefore our "ace in the hole" in terms of delivering the goods for the grant from the Open Society Institute. In his mid-40s, with a gray-flecked beard, piercing hazel eyes, and a stern countenance, André sprang from the Haitian intelligentsia. His grandfather had been a minister of justice. He himself was an intellectual— "always reading and writing" his mother would say—and a lifelong student of Haiti's history. Intense, focused, single-minded, with the goal of creating Haitian generalist doctors with broad-based skills, he had committed to spending six months each year in Haiti, making sure the program took hold.

Establishing a new discipline such as family medicine in a resource-poor country like Haiti is an arduous task. In fact, it had never been done before, at least not in a country as poor as Haiti. First André had to negotiate an agreement—a "convention"—with the ministry of health to make sure the project would continue even if the government changed. Then he had to recruit faculty members from the disciplines of general medicine, pediatrics, and obstetrics and bring them to Miami to learn to be family doctors, there being no precedent in the country. With Michel's help, he drafted the curriculum and schedule. Finally, he needed to recruit five candidates into the residency program each year.

It took a good three years for Haiti's ministry of health to understand the significance of our training program. By that time our original grant from the Open Society Institute had run out and

Medishare had to pick up the costs. The family practice center—the heart of the program—had been forced to operate in renovated space in a building owned by another ministry. Plans to build a new center were put on hold, as it was all Medishare could do just to maintain the faculty.

To understand how revolutionary our program really was, one first has to first understand the conditions at the hospital prior to our program. Hôpital Justinien was large, with open-air wards for pediatrics, obstetrics, general medicine, and surgery, each ward having enough beds for 40 to 60 patients. Nestled between the four main buildings were two completely inadequate buildings that housed the emergency room and the outpatient clinic. The entire complex was painted white with green trim—the colors of hope. Situated on a slight rise on the western edge of the historic district, it was designed to catch the trade winds that blew off the harbor—a kind of natural air conditioning.

Most patients were admitted through the emergency room. Patients admitted "after hours" would have to wait for the doctors to come in the following morning before orders would be written to commence care. The medical staff—private doctors in practice in Cap Haitien—were paid a small amount by the hospital to visit each day, usually two hours or less, to provide care for the poor. The hospital had no functioning laboratory or x-ray unit. Electricity was available only two to four hours a day.

Our program brought residents to cover the major services 24 hours a day. Furthermore, it provided faculty and residents with an outpatient setting to take care of small problems before they became disastrous and to address screening, prevention, and family planning. These changes, particularly the innovation of full-time faculty seeing patients and supervising residents, were at the heart of our revolution. Other changes were more subtle but no less significant. For example, we built, staffed, and stocked a pharmacy within our family practice center, where our patients could get their medicines at nominal costs. We also created a medical record that documented preventive and psychosocial issues and a curriculum that prepared trainees for most of Haiti's health needs, including mental health.

On the wall in the preceptor's room, André hung his motto: "Respect, Compassion, Excellence." It was a revolutionary motto. Prior to our program, social class dominated the profession of medicine in Haiti, as it did in so many other spheres of Haitian life. Medical education was free, but to enter medical school you had to pass an exam. This meant that, in effect, only the sons and daughters of the elite, who could afford a private education, could compete for entry into the profession. After completing traditional residencies in which students gained experience by treating poor patients with no real supervision, they would either enter private practice or emigrate to the United States or France. Never before in Haitian medical education had faculty and residents come together to serve the people.

Poor places like Haiti simply can't afford the luxury of one doctor for children, one for pregnant women, and one for adults, so family doctors were a critical need. It seemed so logical in theory, as Michel and I crafted the grant that got the program started. Three years into the program, the reality was even more compelling—residents and faculty working at the family practice center shoulder to shoulder, changing a 200-year tradition of learning at the expense of the poor into one of learning in order to serve them.

The problem was sustainable financing. We had made believers of the folks at the Open Society Institute, our original funders, and they were willing to help us establish other training programs if the first program achieved financial independence. But the Haitian ministry of health had no money and other possible funders had "strings." For instance, the only funds available from U.S. Agency for International Development were tied to a program to prevent HIV transmission from mothers to their children. United Nations Children's Education Fund had specific priorities based on a national plan designed with the ministry that focused on nutrition, vitamin supplements, and immunization. Unfortunately, the argument that all of these disease-specific goals could be more easily addressed through the training of Haitian generalist physicians fell on deaf ears.

Therefore, the onus for funding fell on Medishare. Medishare had accomplished great things in Haiti, but as a small "niche" charity, major fundraising had always been its Achilles' heel. It was also stretched thin by its commitments to Thomonde and volunteer trips. Medishare's board was growing restive with an open-ended commitment to the residency program and no end in sight. Other long-term sources of funding would have to be found.

The Reverse Zombie Curse

December 2003.   IN THE BATTLE FOR HAITI'S HEALTH, Medishare was forced to open a second front. For years, we had been content to maintain a very low profile toward the medical school's administration, quietly building our programs and developing support among our students and other faculty. But Medishare, by itself, had come about is far as it could. To go further, we'd need the complete support of the medical school. Unfortunately, Haiti's bad press, coupled with the looming malpractice crisis in the United States, had convinced some—most notably the provost of the university and the general counsel—that the faculty and students of the University of Miami had no business volunteering in Haiti.

Malpractice insurance costs, rising everywhere in the United States, hit Miami particularly hard. The administration's concerns were that some poor Haitian peasant would identify the University as a "deep pocket" or that a student would get sick or injured and that the University would be held responsible. These days even a school of medicine had to operate first and foremost as a business.

So a sort of chess game unfolded in which both sides made up the rules as we went along. In this makeshift game the object was to capture the support of the dean of the school of medicine. The administration's opening move was to insist that we offer the students traveler's insurance and that the students sign a waiver of responsibility before going to Haiti. For our part we strove to gain academic credibility for Medishare. Our students returned from each trip with glowing reports of how much they had learned. Furthermore, we also published an article in *Academic Medicine*, detailing Medishare's educational merits. Our students published and presented at international health conferences their experiences screening for tuberculosis among orphans and their case series of children with congenital rubella.

Our grant from the Open Society Institute to establish family medicine residency training in Haiti furthered our academic credibility enormously. No one at our school had ever before received a grant from Mr. Soros's foundation. Plus, as luck would have it, Ellen, our benefactrice at the Open Society Institute, happened to be very good friends with the University's new president, Donna Shalala. President Shalala, a former Peace Corps volunteer and a former Secretary of Health and Human Services, was on a mission to make us a great university. If we could win her over, we'd be poised for checkmate. What won the day, finally, was continuing support from the Green Family Foundation.

The Green Family Foundation was established by a wealthy Miami businessman and former ambassador who gives a healthy portion of his income to charity every year. He also sits on the University of Miami's Board of Trustees. Kimberly, his daughter, runs the foundation. Her faintly counterculture appearance—she's always dressed in peasant blouses and bell-bottom jeans—belies focus and good instincts, which she applies with a hands-on approach to the family's charity. I had met her at Barth's office in an introductory meeting arranged by the Development Office. In attendance were several wealthy women, most of whom were only mildly interested in what we were doing in Haiti. Kimberly, on the other hand, immediately saw the potential to do something special there. Her foundation was focusing on AIDS in children. It had already built a wing for children with AIDS at a hospital in New York City and was about to make a major commitment to our Department of Pediatrics Division of Infectious Diseases. At the meeting I suggested she broaden her scope to include all infectious diseases that were killing children in countries like Haiti and invited her to come with Medishare on our next trip, to see what we were doing in Thomonde and Cap Haitien.

Kimberly instantly fell in love with Haiti. She met Paul in Cange, and our wonderful Thomondois welcomed her as if she were long-lost family. She witnessed firsthand our students, faculty, and their health fairs. She bought into Medishare's vision of community health and community development in Thomonde. Her foundation then funded our community health program in Thomonde and produced a Medishare promotional video and a documentary on Haiti and its health needs to help raise funds for our project. Fortunately, Kimberly's father had picked up on the tension between Medishare and our school. He volunteered to fund a trip to Haiti to see what Medishare and his foundation were doing firsthand. Guests would include President Shalala and the dean. In effect, he had made the dean an offer he could not refuse.

"I'm not happy about this trip," groused the dean during a preparatory meeting about two weeks before our departure. "It's too much a public relations thing. The president loves that sort of thing—I hate it."

Barth, Michel, and I were taken aback by his comment, looked at each other and then at the dean. Perhaps we had overplayed our hand. For myself I saw this as another example of the curse afflicting the school's administration—the same curse of cautiousness and conformity that had led to Joseph's expulsion.

"It won't be political," I promised. "We'll be out in the countryside, with the people, not the politicians. We'll set up some patients for you to see." After the meeting Michel confided that he thought it was a good idea to let the dean see a few patients.

"He's in such a cage d'or," exclaimed Michel. "He never gets to

University of Miami's President Donna Shalala is welcomed in Cange.

see patients anymore. His whole world bounces from one financial crisis to the next. Don't worry, Art. The trip to Haiti will do him good!"

The charter plane from Miami to Port-au-Prince was an hour and a half late. I later discovered it was missing a life raft and had to fly to Palm Beach to retrieve one before it could depart from Miami. My patience was tested as I waited on the observation deck of Aeroport International. Since first proposed by Ambassador Green, the entourage had grown to include Kimberly, President Shalala, the dean, Barth, Ellen, the dean of the School of Nursing, Michel, and another prominent Haitian-American physician, Rudy. Medishare's executive director (also named Ellen), André, and I had gone down in advance to make sure everything was in order. The itinerary included helicopter rides to Cange and Thomonde, a reception in Port-au-Prince, and a visit to our residency program in Cap Haitien.

Those of us on the observation deck waived as each member of the entourage deplaned. *Thank goodness, they're here now. Haiti, cast your magic spell*, I thought, *the same kind you cast on me nine years ago. A reverse zombie curse—one that will liberate rather than enslave, enliven rather than zombify.*

We boarded two prearranged helicopters. I climbed into the smaller one with the president, Kimberly, and a few others, while the dean and everyone else climbed into a larger, open, Vietnam-era chopper. After the crew secured everyone in their seats, we slowly, imperceptibly lifted off the tarmac. It was a clear morning with unlimited visibility. Shortly after takeoff, the president nodded off. *She must have gotten up at 4:00 a.m. Not the kind of spell I had hoped for,* I thought. Looking over to the other helicopter, though, I could see the dean leaning out, peering intently at the countryside below. We had to pass over two mountain ranges and the Artibonite River valley to get to our first stop, Paul's hospital in Cange. A trip that takes us three hours by car took only 30 minutes. Everywhere below us were the small plots of land and tiny homes connected by the footpaths that define rural Haiti. I nudged the president as we crossed Lac Péligre, the lake created by the dam made so infamous by Paul's writings.

From the air, *Zanmi Lasante* has a castlelike appearance. The helicopters made a seemingly miraculous descent to land side by side in the hospital's courtyard. We disembarked, made introductions and were greeted by the Cange brass band, flowers for the president, and a large University of Miami banner that brought surprised grins to everyone's faces. School children seemed to be everywhere, smiling, waving, singing, and lilting "*Bonjour.*" We met with the medical staff and took an abbreviated tour of the hospital. We wanted the president and dean to see what might be possible when a great university partners with a charitable foundation and a community in need. "See what Harvard did!" I deadpanned. "They only have a five-year head start on us!" We then climbed back into the helicopters for the short trip to Thomonde.

Thomonde turned out in force to greet us, waiting patiently as we landed on the soccer field next to the national school. The crowd spilled out of the grandstand built in our honor and encircled the sidelines. Delva was our official greeter, resplendent in a freshly pressed long-sleeved white shirt and a red-patterned tie. Signs and banners on the main street also welcomed us as we made our way to the dispensary and day hospital that Medishare had renovated two years before.

Hundreds of patients waited there—to see not us but the Haitian doctor Medishare supported. It was a typical "day in the life" of the clinic since Medishare brought health care back to Thomonde: women receiving prenatal care and family planning, children with infections, old folks with innumerable afflictions.

Our "day hospital," originally intended as a triage station for patients who might need to go to Cange, was obsolete and overwhelmed the day it opened. One of our hopes for the trip was to generate interest in building a real hospital and a larger clinic on land donated by Thomonde. Both the president and the dean were moved by the obvious suffering of the eight patients in the day hospital. We had only six beds, so two patients were resting on mats on the floor. I introduced the dean to one young patient, who was swollen almost beyond recognition, a condition caused when the kidneys leak protein.

"This is a cascade of catastrophes, John," I said to the dean, after I explained to the patient in Creole why we were visiting.

"What do you mean?"

"It most likely started with a mosquito bite."

"A mosquito bite?" The dean looked puzzled.

"Yes. Lots of mosquito bites here, as there's no mosquito control program. And the kids scratch them and inoculate strep under the skin, which leads to cellulitis and impetigo. Some of the strep are so-called nephritogenic strains and that leads to post-streptococcal glomerulonephritis, nephrotic syndrome, and kidney failure."

The dean gave the patient a silent look of empathy and compassion. As we were leaving the dispensary he said to me, "I want to come back in the spring and just see patients."

"That would be great, John!" I responded. "You have to come

here and see it to truly understand the need and the difference we can make."

He put his hand on my shoulder: "You're right. You have to see it to understand!"

Haiti, the land of contrasts, was working its miracles. A minor miracle came at our next stop—a plot of land and a shell of a building on which we hoped to build a new hospital and clinic for Thomonde. After Delva explained what we hoped to accomplish, the president spoke spontaneously about the commitment the university would make to help. I listened incredulously, my hopes and dreams for Thomonde expressed by the president of the university. It was as if my words were coming out of her mouth! For the rest of the visit to Thomonde she alternated between praising our health care workers and brainstorming with various members of the entourage as to how we could get more students and faculty and other schools involved, and how we could write grants and raise funds.

Of course, buy-in by the school and university would be a hollow victory if we weren't also winning the war for the health of Thomonde. We stopped at our guest house as Marie was wrapping up the final day of the Thomonde census. Marie is a University of

Thomonde's community health workers.

Miami nurse who used to work for Margaret as an AIDS research nurse. After hearing about our program, she decided to give up her work in Miami and return to Haiti as the in-country coordinator of the Green Family Foundation Initiative. With her guidance, our small army of accompagnateurs (community health workers) had visited every home in the district and cataloged each family's health issues.

This gave us, for the first time, accurate information on all the health problems in the commune. In the process, our accompagnateurs established a direct, personal connection with every family in Thomonde. What seemed like an impossible task had been made simple: Fifty community health workers serving 50,000 people meant that each worker served 1,000 people. Since the average Haitian family had five children, that meant each worker was responsible for fewer than 200 households—a remarkably manageable number.

The president asked if there were any issues with the traditional healers. "Not at all," Marie explained. "First, everyone knows everyone else. We're all friends. Second, we had classes and seminars that included, in addition to the community health workers, the *fanm saj*, *dokte fè*, *bokars*, and *mambos*. So everyone sorted out who was responsible for what, and we refer patients to each other. For example, if a patient is lovesick, they'd be referred to the *mambo*. Fever and weight loss—that's our province."

Next, we all trudged down a footpath to the house of Sonson. Sonson, his wife Charitable, and their seven children were all enrolled in our direct observed therapy program. When diagnosed with tuberculosis six months before, Charitable was wasted and too weak to work—a zombie, if you will. Sonson practically said as much. In discussions with his neighbor, Isoni, he attributed his family's misfortunes to a curse placed on them by an enemy who wanted their possessions when they were all dead.

Isoni was more than a neighbor, however. She was also one of our accompagnateurs. Slowly, over time, in the course of daily con-

The University of Miami team visits Sonson at his home.

versations and visits, Isoni convinced Sonson that, in Charitable's case, the Voodoo charms weren't working. What did they have to lose by consulting the Medishare doctor? By that time, Charitable was too weak to walk and Sonson was too weak to carry her. Fortunately, a neighbor lent them his mule to make the one-mile journey to our clinic. Now, thanks to the daily visits of our health workers, he and his family were healthy again. He was working in his fields. His children were back in school. It seemed the curse had been reversed.

While I must relate that I've never actually seen a real zombie in Haiti, I know several credible Haitian doctors who do believe they exist. Personally, I'm skeptical. If zombies do exist, it's probably a rare occurrence for a curse to actually be placed. Just the threat of one assures conformity to society's norms. But I've probably seen thousands of patients, just like Sonson, who, in searching for meaning in their fate, thought a curse had been placed on them. Treatment gives hope, and hope is the antidote—"*Lespwa fè viv!*" ("Hope makes us live!").

We now have hundreds of patients like Sonson in Thomonde under treatment or who have completed treatment for tuberculosis and AIDS. After two decades of being outsmarted by a small bundle of nucleic acid and a protein coat, we're finally figuring out how to get the upper hand. The people of Thomonde, this most improbable place, have shown the way.

As our guest house was now serving as headquarters for the community health workers and as a residence for Marie, Delva, on his own initiative, had started a new guest house. When finished, it will accommodate 38 people, two per room, each room with a shower and toilet. Delva even included a courtyard and fountain in the design. Nimi, who led the group of women who did all the cooking and cleaning during our visits, had prepared a cornucopia of Creole delights for our lunch—macaroni and cheese pie, *piklis* (a spicy coleslaw), plantains, rice with beans, a beet salad, and more. The dean, the president, and the rest of the entourage filled their plates and marveled at the skills of Nimi and her crew.

While everyone was eating, Delva and I slipped away to a quiet corner. Sitting side by side, we each sipped a Prestige, the Haitian national beer. We congratulated each other on our achievement: *"Enkwayab, Delva. Prezidan ak dwayen Inivèsite Miami isit nan Thomonde. Vrèman, lwa ou-yo travay pou nou jodi-a!"* ("Amazing, Delva. The president and dean of the University of Miami here in Thomonde. You sure have the spirits working for you today!"). Delva and I were holding hands. In Haiti you can hold the hand of a friend of the same sex. It's simply a symbol of friendship. We continued our discussion in Creole.

"What do we do next, Dr. Fournier?"

"We'll have a hospital and a new dispensary. Maybe four or five doctors and eight or nine nurses. I'd like to see Thomonde as a rural training center, a place where doctors can learn to work with nurses and community health workers. Your new guest house would be perfect."

"The University of Thomonde!"

"We could do that," said Delva. "Good idea!"

"So, which *Lwa* did you pray to to get Medishare to Thomonde?" I asked with a knowing smile.

Delva returned a smile of his own. "Dr. Fournier, do you even have to ask?"

"My guess would be Damballah, but I really need to know. You never know when I might need some help!"

The helicopter ride back to Port-au-Prince was the most visually beautiful experience in my life. The Artibonite shone as a silver ribbon stretching to the horizon. The setting sun illuminated the hills and mountains in russet and gold. They in turn cast gray-green shadows across the valleys and plains. Waves of gold and shadow—a metaphor for Haiti, land of contrasts. Beauty and misery, slavery and freedom, death and resurrection. What more could one ask for from a country?

Conundrum

THE NEXT DAY WE WERE STANDING in the general airport, waiting for the Caribintair clerk to write out a receipt by hand. We had just flown back from Cap Haitien, where the dean and the president had toured the family medicine residency program we had founded with the assistance of the Open Society Institute. I was escorting them back to the international airport so that they could return to Miami for a faculty dinner the president was hosting that evening. It seemed like a good time to share my vision of a Center for Humanitarian Medicine.

The dean had opened the door by mentioning to the president how we had struggled for a while in defining the relationship between Medishare, our charity, and the school. In an era of falling income, rising costs, and high liability concerns, some questioned whether there was any legitimate reason for our school's faculty to volunteer in Haiti. The sentiment seemed to be, "Let Medishare do it, and keep the school out of it!" That's where the Center for Humanitarian Medicine came in.

I had come to believe, over several years, that a "perfect storm" had been brewing, threatening to blow away all that was noble in the profession of medicine. The convergence of forces that formed this tempest included the corporatization of medicine, the rising debt of medical students, the falling incomes of physicians, the malpractice crisis, and lifestyle choices—choosing careers that don't involve nightcall, long hours, hard work, and low remuneration. In 2004 only three of our 150 graduates had chosen careers in family medicine. In a world where medicine had become a business and health care a commodity, the committed, compassionate, caring physician who treated everyone regardless of their ability to pay had become an endangered species.

Through the Medishare years I've loved how our students have embraced the humanitarian cause during their trips to Haiti. Responsibility is an even greater teacher than experience. In Haiti they got a heavy dose of both. Michel and I would marvel at how quickly they ascended the learning curve and laughed secretively among ourselves. "They learn so much, and they don't even know they're learning!" We had created, without knowing it, the anticurriculum—no tests, no grades, no rote memorization of obscure minutiae, just the pure joy of learning what you need to know to actually help people. But the sad truth was, once they returned to an educational environment that's become increasingly dehumanizing, only a few were able to maintain their commitment.

A Center for Humanitarian Medicine would bring together faculty from several disciplines to teach and role model professional values. In the process they could provide services to those in need both in Miami and abroad and demonstrate creative ways to alleviate suffering regardless of their patients' lot in life. It seemed like a good idea to me.

" 'Humanitarian' has become a bad word, Art," responded Ellen. "You definitely don't want to call it the Center for Humanitarian Medicine."

"Humanitarianism is a bad thing? When did that happen?" I protested, crestfallen.

"You need to read David Reiff's *A Bed for the Night*. It's depressing but true how humanitarians like yourselves are taken advantage of by the realities of politics."

"But, you see, we've got a huge need in medical education these days to shore up professional values."

"I understand the concept and support it. I'd just call it something different. How about something like what Paul has created at Harvard, a Center for Health and Social Justice?"

"That would never fly in Miami. It sounds too much like socialized medicine. Besides, I'm coming more from my role as an educator and what we need to be teaching our students. Paul's coming from his commitment to service."

"Ellen's right," chimed in the president. Why not call it the Center for Human Security—a takeoff on the Department of Homeland Security!"

This conversation was a mixed blessing—perhaps I was naive in my beatification of the need for humanitarian values in medicine. I could probably use an upgrade in sophistication on humanitarianism, and perhaps Mr. Reiff's book could serve as a useful primer. On the other hand, Ellen and the president understood the concept of a Center for Humanitarian Medicine and agreed with the need. I took this as a green light to proceed. We continued the discussion during the 15-minute cab ride from the general airport to the international airport. A short while later, I escorted them through customs and waved goodbye as they boarded for Miami. The trip had gone better than I had hoped. Haiti had worked its magic charm. The president and the dean were committed to our cause. But where would it all lead next?

I bought *A Bed for the Night* shortly after I returned home. Ellen's assessment was accurate. The book was depressing, irrefutable in its logic, and offered no solutions. The author starts with the trap of neutrality that has entangled the International Red Cross, forcing it unintentionally to enable some really bad characters. He then moves on to the abuse of humanitarianism in Bosnia, Kosovo, and Rwanda.

His point, hammered home over and over again, is that humanitarian aid is a cosmetic Band-Aid on the really horrible things that some human beings do to others.

Shortly after we returned, Haiti erupted into chaos. It was no secret that forces inside and outside Haiti wanted to get rid of President Aristide. Rebels in new uniforms, well armed, and rumored to be supported by the Central Intelligence Agency surfaced, first in Gönaives, then in Cap Haitien, and finally in the central plateau. Many were former army veterans hoping to restore the armed forces banished by Aristide. Several of the rebel leaders had a history of human rights violations during the Duvalier era. They had been hanging out on the other side of the Dominican border, making occasional forays into the Central Plateau. They had passed through Thomonde several months before, frightening the Thomondois. Worse yet, they stopped some *Zanmi Lasante* volunteers traveling from Cange to Las Cahobas, forced them to abandon their supplies and vehicle, and later killed two security guards at the dam at Lac Péligre.

Angelik

March 2005.   "Not a good site for a health fair," murmured Ellen, Medishare's executive director, as we climbed out of our trucks and descended down a barren ridge to the school at Croix Rondo (Rounded Cross). "Too much wind! Too much dust!"

The school was a large open cement platform with a few upright posts that supported a tin roof. At least we'd have some shade. Our health agents had tied ropes and sheets to some of these posts and squared off a corner with portable blackboards to provide one small area of privacy for prenatal care. The sheets billowed and popped in the wind like poorly trimmed spinnakers, adding to the noise created by the din of the crowd and the wind howling under the roof. There were only three trees on the otherwise barren ridge and little grass. There had been no rain in five months. From the ridge one could look out over what seemed like the entire Plateau Central, extending to the horizon. The trade winds, blowing from the northeast, raced across the plateau until they reached the ridge, scouring us with blasts of hot air, powder, and dust that peppered our skin, stung our eyes, and clogged our nostrils. Crowds had gathered in the sparse shade of the three trees where our health agents had set up tables to register patients in advance. I estimated 300 patients were waiting to see us. It was going to be a long day.

My team included several returning Medishare veterans, most notably Joseph, about to finish his second year in our family medicine residency program; Rachel, a graduating senior who had come on three prior trips; and Rick, who had come four times before and was now an emergency medicine resident. Nine enthusiastic medical students, a volunteer nurse, and four Haitian health workers rounded out the group. The health fair the day before at Boucantis, an even more desolate and remote part of the commune, had gone very well. Construction of the new hospital/clinic complex in Thomonde was proceeding more quickly than anticipated, and we had just received some good news about funding for our nutrition program. Considering all this, I should have been in better spirits. But our trip to Thomonde had been delayed a day by violence. On the day we arrived, the United Nations forces had engaged in skirmishes with former rebels in Petit Goave, with casualities on both sides, and demonstrations had erupted in Terrier Rouge. Terrier Rouge straddled our usual route up Morne Kabrit on Route Nationale 3. So after a night in Petionville, we were forced to take an alternate route to Thomonde—an exhausting eight-hour journey through St. Marc and the Artibonite Valley. Plus, I had a mild knee injury from a previous trip, which was aggravated by long car trips or standing for long periods of time. Between the ride up, the ride out to Boucantis the day before, five hours on my feet there, and a one-hour ride to Croix Rondo that morning, my knee was already throbbing like a toothache. The pain, fatigue, dust, and noise combined to create an unusual sense of foreboding.

The boundaries of the commune of Thomonde were defined politically, not geographically. To get to Croix Rondo, our convoy had to first ascend and then descend the Thomonde caldera's volcanic lip, cross a river, climb back up onto the central plateau, turn off a bad road onto a glorified footpath, and bounce and pitch for another half hour to the ridge where we would hold our fair. I estimated it would take four hours to get from there to our clinic in Thomonde on foot and two hours by mule. Because of its isolation, Croix Rondo had yet to have a health fair.

Five months without rain had made even the usually oasis-like Thomonde basin look dry. On the central plateau, on the other side of the Thomonde rim, things were positively parched. In over 100 visits, we had witnessed floods, revolution, and the effects of crushing poverty, but I was mentally unprepared for drought and its sister scourge, famine. The media, preoccupied as always with the politics and violence of Port-au-Prince, had never mentioned either. More than disease, rebellion, or flooding, crop failures associated with drought could easily reverse the progress we were making in Thomonde. And unlike disease, I had no answer for it. It was not as if we could treat thirst with antibiotics!

We would see more sickness, particularly among the children, at Croix Rondo than we'd seen in Thomonde proper. Several children with severe malnutrition stood out in the crowd, easily identified by their telltale red hair and swollen bellies.

Because it was Croix Rondo's first health fair, it took us longer than usual to get set up and organized. Pregnant women were seated outside the billowing sheets that defined the prenatal care station. Rick and some students arranged some handmade school benches into a child care section next to them, and Joseph structured a similar station for the sick adults at the far corner of the school. As soon as they were registered, our patients climbed about a hundred yards up from the shade of the three trees and waited patiently in the shade of the north side of the school to be called to a station. Most were dressed in their Sunday finest, sadly powdered with dust on the side that faced the wind. I floated from station to station, encouraging my students to be thorough but efficient, talking to the patients in Creole, and writing brief notes on their charts to hasten the process of evaluation and referral. By 11:30 the health fair was in high gear.

"Reynaud!" I yelled at the top of my lungs, attempting to get the attention of our health agent. "*Vini isi!*" ("Come here!"). "*Nou genyen lòt timounn ak tibèkiloz!*" ("We've got another child with tuberculosis!"). Kea, a first-year student, had just correctly diagnosed tuberculosis in a five-year-old boy by feeling the telltale matted lymph nodes in the back of the child's neck. It was our third case of TB that morning. "This particular way that TB presents is called 'scrofula,'" I explained to Kea, while congratulating her on her clinical acumen.

Kea burst into a radiant smile and then a sigh. "Gee, I feel like a doctor!" she gushed.

"Yes, before Medishare, that child and countless like him would have died. Now Reynaud will make sure that he's on treatment no later than tomorrow. His whole family will be tested. You've made a difference. It's a great feeling. Enjoy it, just for a moment, before you see your next patient," I counseled.

My optimism was starting to return. In addition to the three children with tuberculosis, we had found one child with pneumonia. All would likely have died had we not come to Croix Rondo that day. Plus, we found one boy with undescended testicles, an old man with cysts in his testicles related to elephantiasis, and an older woman with a large neck mass. These patients could all be helped by the surgical team that Medishare was sending down in June. Just then Rick touched my arm from behind to get my attention. "Dr. Fournier, I need you to see this child. Please tell me that what I'm seeing is not what I think it is."

"I'll be right there, Rick. Let me just finish with Kea and Reynaud." I joined Rick at his "kidscare" station a few minutes later. He spoke as softly as he could and still be heard above the wind and the crowd, "This is Angelik. She's 9. Her mother brought her because she's got marks on her genitalia. I think its HPV."

HPV is short for human papilloma virus—another smart virus— the one that causes cervical cancer. HPV in this particular anatomic location, in this child, could mean only one thing—sexual abuse. I picked Angelik up and carried her behind the sheets to the prenatal care station. Fortunately, the prenatal team was between patients and

Young boys in Thomonde with their homemade "futbol." Photo by Wassim Serhan.

the examining table was empty. Rick and Angelik's mother followed me. "The marks are a curse," she whispered to me, careful that the other patients would not hear.

Angelik was a beautiful child with dark, unblemished skin, large eyes, and neatly braided hair capped with little white beads. Her face, however, was frozen in fear. It was impossible to know whether her fear was caused by the trauma she might have experienced, the possibility that she was cursed, or all the attention she was getting from these funny-looking strangers. "Don't be afraid and don't be ashamed," I whispered into her ear in Creole. "I need to look down here to see what these 'marks' are." I gently positioned her on the table, bent her legs, and removed her panties. There was no mistaking the classic lesions of HPV that covered the top of Angelik's vulva and extended down, in clusters, over her clitoris and inner lips. Rick, who had put on examining gloves, shined his penlight on her vagina. "There's no hymen and the mucosa looks inflamed," he murmured. I quickly peeked at where his light was shining and confirmed his findings.

"It's a curse!" Angelik's mom exclaimed again. The marks she was referring to are commonly known in English as "lover's warts." This seemed a particularly cruel appellation given the circumstances and I knew of no equivalent Creole translation. How could I explain this in terms that Angelik and her mother would understand?

Having seen tens of thousands of them in my career, there was simply no alternative explanation. So there it was: the smoking gun! Ever since my first trip to Haiti, when Regine had stroked Ruth's leg and asked for her wallet, I had recognized the relationship between the exploitation of women and children and the spread of sexually transmitted infections. I had written about it and spoken passionately about it. But seeing it, literally staring me in the face, shocked and startled me. Evil existed in the world, even in my beloved Thomonde. "Excuse me," I said in both English and Creole. "She can get dressed now."

"Rick, I'll be right back. We need a plan." I exited the curtain and walked over to our trucks, in part to talk with Marie, our nurse coordinator, and in part to gather my thoughts. Marie was reclining in one of the seats, catching her breathing in front of the car's air-conditioner. Usually an inexhaustible bundle of energy, Marie had suffered an asthma attack, precipitated by the dust, forcing her to retreat to the relatively pure air of one of our cars. "I've got a little girl who's been sexually abused," I said solemnly.

"What can we do?"

Perhaps it was fatigue brought on by her attack, but Marie seemed oddly unsurprised by my news. It was as if, with all the health fairs we had been through together, it was only a matter of time before we had a patient like this. "We can tell the police, but they rarely intervene in cases like this. For all we know, they might even be the problem! The best thing is to have her mom bring her to see our nurse-midwife in Thomonde. At least she's a woman and won't add further to the fear and blame."

I returned to the health fair. Before speaking to Rick or Angelik's mom, I locked in my eyes to the child's eyes. For the first time in my career I dropped all pretense of objectivity, that cold distancing mechanism I had used to protect myself from engagement with Regis, Marc, Previlus, Fanesse, and the others—the "hey, I'm sorry, but you're the one with the disease" attitude that seems to afflict so many doctors. For a long silent moment we looked at each other without hierarchy, not as doctor and patient, adult and child, Haitian and American, but simply as two human beings, one suffering and one wanting only to help: *"Nou renmen ou. Nou kapab ede ou, nou kapab retire mak-yo"* ("We love you and we can help you. We can make the marks go away"), I repeated, three times, almost like a religious chant, as I held her jaw in my hand to keep her eyes focused on mine. By the third incantation I managed to elicit a soft smile. I returned her to her mother's arms.

Angelik's mother was dressed in an orange-yellow blouse with shoulder pads and a flowered skirt. She looked about 35 years old, but her harsh countenance, probably reflecting both fear and concern for her daughter, may have made her appear older than she actually was. She knew exactly what I was thinking and half-heartedly repeated her assertion that the marks on her daughter were a curse. "It's not a curse. It's an infection, nothing more." I responded. "We can get rid of it. Don't blame yourself and don't blame Angelik."

"*Dakò*," she agreed.

"I have to ask you some more questions. Are there boys in the house?"

"No. She only has sisters," she answered emphatically.

"Could it have been her father?" I asked straightforwardly.

She answered with an equally emphatic "No!"

"Did Angelik say anything about who might have done this?"

"No. She has not spoken in weeks."

"Well, she needs all of our love and you need to protect her. Make sure there's another woman around when you go to the market. Bring her to see the nurse-midwife at Thomonde. A simple treatment will make the marks go away—a purple potion. ("Podophyllin!" I said as an aside to Rick). "When the marks are gone, if you treat her as always, she should start talking again." With that her countenance softened and she gathered up her things to leave. I took Angelik by the shoulders and stooped to kiss her on the forehead. *"Bondye beni ou"* ("God bless you"), I spoke as I engaged her eyes one more time. "Rick, I've already told Marie about this. Make sure that Reynaud knows and brings her to Thomonde tomorrow. She's going to need tests for chlamydia, gonorrhea, syphilis, and HIV, in addition to the podophyllin treatment."

"What do you suppose happened and who did this to her?" asked Rick when they had gone.

"We'll never know," I responded. "It's possible she wasn't raped but only molested. The role of digital intercourse in the transmission of HPV is poorly understood." My wall of objectivity had reascended. I had become a professor again. "Your findings on penlight exam were ominous, though. I fear the worst. As to who did it, in the States, it would most likely be a family member or friend. Here, you can't assume that. While it's true that Haitian girls start having sex at a young age, it's usually after puberty and consensually, as part of courtship. Remember, there's a miniwar going on out there. We're only a couple of miles from the Dominican border and both rebels and United Nations troops are combing these hills. Soldiers are notorious for taking advantage of the powerless and even the UN troops are not above suspicion. Did you read that article last week about rapes in the Congo? A child Angelik's age would be pretty much on her own on the days when her mom was at market, particularly if she wasn't in school. It may not have been directly forced sex; it may have been pseudoconsensual. There's a drought here. Delva tells me the wells are going dry, and he's had to truck in water! Angelik may have let someone have his way with her for something as basic as a drink of water. Let's get back to work. We can talk about this more this evening."

By the time we arrived back at the guest house my knee was really throbbing. I asked Delva for some rum, which eased the throbbing and numbed my brain. "You didn't tell me just how bad the drought was, Delva," I admonished. "It was brutal out there at Croix Rondo."

"I'm sorry, Dr. Fournier, but I did tell you. We're lucky here in Thomonde. At least there's still water in the river. But it's like Croix Rondo throughout the Plateau Central," replied Delva. "Remember, I told you we were trucking water to the outskirts of the district. You just had to see it with your own eyes."

I took a brief shower, being careful to conserve water and then took a nap before dinner. Rick did not bring up Angelik for discussion after dinner. Perhaps I had sufficiently addressed his questions or perhaps he felt her privacy had already been violated.

The next morning I had to leave for a meeting in Port-au-Prince. As the truck rumbled past the dispensary, I peered into the shadows of the waiting area. It was too dark in the dispensary to discern individual faces, but I could clearly make out Angelik's mom's orange-yellow shirt with shoulder pads and her flowered skirt. A small girl's form in a white communion dress and white hair beads also radiated from the darkness. Well, the system worked. *The health worker got her here. She'll see the nurse-midwife,* I thought to myself, as the dust raised by my truck obscured them from view.

It's a start. Please, God, let her heal. Just the possibility of an innocent child trading sex for something as basic as water overwhelmed my consciousness. *And send us rain!* I thought, as we headed out of town. Then, remembering the Gonaives floods and the old admonition about being careful what you wish for, I added an addendum to my prayer—*just not too much!*

Bicentennial

THE YEAR 2004 WAS SUPPOSED to be a festival year in Haiti—its bicentennial, marking 200 years since its independence from France. The year didn't turn out that way, however. In the spring of 2004, the U.S. media presented the growing rebellion in a favorable light and painted the Haitian government under Aristide as a "failed state," incompetent at best, corrupt and evil at worst. Aristide, for his part, seemed to be sleepwalking through the crisis. I wondered how many of his nine political lives he had left. *Perhaps someone put a zombie curse on him,* I thought. Certainly he had made mistakes during his administration. I had become disillusioned with his political party, Lavalas, after Delva lost the Thomonde mayoral election. I had no empathy for the rebels, however. They were clearly mercenaries with a sordid past. Worse, the theatrical violence I had noticed early on as an integral part of the Haitian political landscape was no longer just for show. Lives were being lost.

Inexorably the rebels marched from the north and the central plateau toward Port-au-Prince. Resistance evaporated in the face of their advance, almost as if the whole rebellion were orchestrated on both sides. When Cap Haitien fell under rebel control, services at our family practice clinic were disrupted for three days. In the Central Plateau our ambulance was "borrowed" by rebel troops three times. Each time, thanks to the consummate negotiating skills of Delva, it was located and returned to Thomonde.

I was shocked to see a picture in the *Miami Herald* of rebels escorting Red Cross troops through Gönaives. The article talked of the "humanitarian crisis" developing in the north as a result of the roads being cut by the rebels and the inability of the Red Cross to deliver emergency food relief. The rebels had let them through in exchange, it seems, for a major photo opportunity. I drafted a protest e-mail to the Red Cross and received a polite letter back, explaining how they could not be effective if they didn't remain neutral. Meanwhile, my Haitian-American friends were advising me to give it up.

"The handwriting is on the wall, Art. The Haitian elite, the U.S. Embassy, and the media are all over Aristide. Remember, after he is gone, you still want to be able to function in the country," they pleaded.

We were forced to cancel our spring volunteer trip. Communication with Thomonde was sporadic. Marie was marooned in Port-au-Prince, our e-mail was down, and Delva could only occasionally get to the capital to send us phone messages. Fortunately, our Haitian infrastructure held together. The hospital in Cange never shut down. The Thomondois protected our guest house and clinic from looting and vandalism. Medical service interruptions were minimal. Through it all, our community health workers slogged it out, assuring that the mundane miracle of direct observed therapy continued.

Eventually, a small contingent of U.S. marines landed. A few days later the United States and France declared they had no confidence in Aristide and, in the middle of the night, escorted him out of the country. He later claimed he was "kidnapped" by U.S. forces. The bizarre history of the relationship between Haiti and the United States had taken a new turn. When a "caretaker" Haitian-American prime minister was flown in from Boca Raton, Florida, he hailed the rebels as "freedom fighters."

A charcoal market, Cap Haitien.

After four months of headlines, Haiti's 15 minutes of fame ended for a while. The U.S. public and media turned their attention to the presidential campaign and the conflict in Iraq. On the ground in Haiti, things in Port-au-Prince were as bad as or worse than when Medishare first came 10 years earlier. There was no electricity, there were huge piles of uncollected garbage, there was political violence and the treasury was bankrupt. United Nations peacekeeping forces seemed reluctant to intervene in the simmering feud between polarized political factions, sure to erupt into violence again the next time an election is attempted.

The political turmoil of the winter and spring was followed by natural disasters during the rainy season. In May, flooding killed hundreds in the towns of Jimani and Fonds Verrettes. In September, floods spurned by tropical storm Jeanne killed thousands and left hundreds of thousands of people homeless in the region around the city of Gonaïves.

The international peacekeeping force and international relief agencies seemed totally unprepared. The United States pledged a paltry $50,000 for humanitarian relief but then increased its pledge to $2 million after it was shamed by more substantive contributions by Venezuela and Cuba. In reality, little aid from any source trickled down to those in need. People who had gone without food and water for days were scattered from food distribution centers by rifle shots and tear gas. Marie, Delva, our doctors in Thomonde, and our partners in Cange packed truck loads of medicines and drove them into the disaster area, donating them to the agencies—CARE and UNICEF—responsible for recovery.

"It was worse than you could possibly imagine," Marie told me. "The stench of death was everywhere. But almost as bad—depressing, really—was how disorganized the international relief effort was. No one knew what the others were doing; no one took charge. Women and children were walking around in a state of shock."

There are those who believe that events such as the floods of Gonaïves are proof that Haiti is, indeed, cursed. There is a legend in Haiti that Toussaint L'Ouverture made a pact with the devil to drive out the French and that all of Haiti's woes spring from that Mephistophelian source. From a theological perspective, the "Haiti is cursed" theory should not be dismissed without careful deliberation. The recurrent flash flooding in Haiti is the result of deforestation of the country. Many Haitians believe that *Lwa*, the Voodoo spirits, live in the trees, particularly *mapou* and mahogany. Being totally dependent on charcoal for fuel, the Haitian peasants are forced to choose between survival and the wrath of the gods.

In truth the last secret of the zombie curse is that Haiti's curses are not supernatural in origin at all, but rather the consequences of the actions of men. I recalled my conversation with the foundress of the Haiti Baptist Mission concerning the mudslides caused by tropical storm Gordon 10 years before. Since then a decade had passed and nothing had been done to address the root causes of this kind of disaster. Certainly the cutting of trees for charcoal in the most densely populated country in the Western Hemisphere established conditions that in a major storm like Jeanne could sweep away all in its path. But here, once again, blaming the victim makes it easy for us to walk away. It's the peasants' fault for cutting down the trees, we assume. The real cause of the death and despair of the floods is the same as the AIDS epidemic—poverty and its consequences. Poverty drove the Haitian peasants to try to terrace and till every inch of marginal farmland. Poverty continues to drive them to use the cheapest fuel possible, charcoal, knowing full well they are killing their country in the process.

If one probes deeper, things become even more complicated. Gonaïves lies on a flat plain at the mouth of the Artibonite River. The Artibonite, Haiti's largest river, starts at the infamous dam at Lac Péligre. Rumor has it that the night of the storm, the dam-keepers, fearing the dam might burst, opened its flood gates, dumping billions of gallons of water down the valley toward the sea. Canals intended to facilitate drainage at the mouth of the river had long been neglected. In practically an instant, man and nature had conspired to transform 250,000 people from a state of decent poverty to one of despair.

Worldwide, in 2004 the AIDS epidemic showed no signs of abating. International agencies and giant foundations gave millions of dollars to make AIDS drugs available at lower costs. While this was undoubtedly a good thing, little was being done to address the root cause—the exploitation of the poor, particularly women. Until the world wakes up to that reality, that little clump of nucleic acid will continue to outsmart us.

In fact, if an evil scientist or dictator were scheming to design a plan to spread AIDS around the world as a weapon of mass destruction, he couldn't come up with a better plan than the way, in the United States and the developing world, we deal with issues surrounding poverty. Foremost is the exploitation of women—always poorer than their male counterparts, dominated by them, kept socially and politically inferior, and forced into such survival choices as prostitution or early marriage. Then there's the problem of political violence—sexual coercion associated with armies, rebels, and police.

Our corrections institutions long ago abandoned any attempt at rehabilitation. They have become revolving doors of drug users, sex offenders, the homeless, and the mentally ill. Prison authorities turn a blind eye to sex, coerced and uncoerced. Some in society even see the rape that invariably occurs in prisons as part of the punishment due a prisoner who's committed a crime. Lack of access to care, the cost of medicines, and inadequate public health and public education are fueling the epidemic. Even immigration policies have helped spread the disease, breaking up stable relationships as one partner or the other is allowed to emigrate legally or flees to work as an illegal alien.

In Miami the war against AIDS has been a virtual stalemate for several years. Newer AIDS medicines dropped the death rates among known AIDS patients for a while, but problems like drug resistance and serious side effects have surfaced. Meanwhile, the number of new cases keeps rising. This should not be surprising. First, the programs funded to care for the poor with HIV emphasize expensive medical treatment in the advanced stages of the disease. There is little attention to prevention or primary care. More importantly, in a country that has yet to acknowledge health care as a right, both the programs and the people who provide the care find it impossible to escape the "zombie curse" of prejudice, blame, and conventional thinking.

For example, a young working mother of two recently walked into our private practice complaining of two weeks of the flu. As soon as I saw her, it was clear she was desperately ill with pneumonia associated with AIDS. As I delved further, I discovered the following: She learned that she had AIDS while pregnant with her second child and had been on highly active antiretroviral therapy until she started working. At that point her Medicaid was cut off. She made too much money! To make matters worse, she was told by her employer that she'd have no health insurance during her probationary period. When she finally did qualify for insurance she was denied coverage for AIDS treatment because her disease was a preexisting condition. During the time she was without insurance and without treatment, her immune system deteriorated from a level robust enough to fight off opportunistic infections to a level almost incompatible with life. She almost died and her children were almost orphaned because she found a job and therefore lost her insurance! The cost of her totally preventable hospitalization was probably 10 times the cost of maintaining her insurance.

Ironically, in 2004, at least with regard to the AIDS epidemic, things in Haiti were a little better. Countrywide, the prevalence had actually fallen a bit, a small blessing that's probably attributable to a falloff in exploitative sexual tourism, new programs in public education, and the superhuman efforts of Paul Farmer, Bill Pape, and others engaged in the AIDS arena. Haiti is a recipient of funds from the presidential initiative to fight AIDS. Wisely, the people at the U.S. Agency for International Development who manage this program in Haiti realize that the war against AIDS can't be fought in a vacuum. They've designed a countrywide multifaceted program that is creating a general health care infrastructure with the capacity for AIDS care, rather than trying to treat only AIDS. Programs in prevention and nutrition as well as AIDS care and treatment are being implemented. Our family medicine training program is playing a role in this effort, not only through its services to patients in Cap Haitien but also by training the next generation of Haitian doctors to be competent in all aspects of AIDS care. In 2004, over 1,000 patients with AIDS joined the 15,000 patients already receiving comprehensive health care services from our residents and faculty in Cap Haitien. However, if civil war erupts, these modest gains will be rapidly reversed by the rape, coercion, disruption of services, and worsening poverty inevitably associated with war.

Actually, even in the absence of overt civil war, the virus is mounting a counteroffensive. Skirmishes between pro- and anti-Aristide forces in the slums of Port-au-Prince have driven the market

Sonson's son after completing Direct Observed Therapy. Photo by Wassim Serhan.

women who used to work there up into the relative safety of
Petionville. An estimated 6,000 people are now "working homeless,"
living in the streets that serve as the outdoor markets of Petionville.
It's the "Petri dish theory" revisited, with every bodily function performed in the streets. Haitians of all walks of life, all levels of society,
refer to this sea of humanity as "*ti-Fallujah.*"

Miraculously, in Thomonde incredible progress has been made,
despite Haiti's precarious political situation. Paul received a large
grant from the Global Fund to Fight AIDS, Tuberculosis, and Malaria, which has made medicines available to all and allowed us to
expand our team of health workers. With treatment and a simple
but effective delivery system (our community health care workers)
the stigma of AIDS and tuberculosis is dissolving. Over 400 patients
have been enrolled in our direct observed therapy program. To date,
not one has died. There are two secrets to our success in Thomonde.

Thomonde's community health nurse.

First, we've trained Thomondois to be peer educators, counselors, and therapists. This has enabled patients to care for themselves and to take an active role in their own health, with the health care workers serving as a safety net if things don't go according to plan.

My colleagues in Miami think I'm joking when I tell them we now have a better and fairer health care system for the poor in Thomonde than we do in Miami, but I'm actually serious when I say this. How ironic that Thomonde—one of the poorest regions of the poorest country in the hemisphere—has a more effective model of care than most of the United States. In the States, frequently those who need health care the most get it the least and those who need it the least get it the most. People of all socioeconomic strata take health care for granted; they just want to take a pill and to have their problems go away. Paradoxically, many of our health problems here in the States are the by-product of our affluence—rampant obesity, even among children, along with its complications, diabetes, hypertension, and heart disease. Even the poor in this country suffer from a relative affluence. For examples, one need look no further than the popularity of cheap fast food and its contribution to the epidemic obesity or the use of cheap crack cocaine and its role in the spread of AIDS.

In Creole there is no verb equivalent to the French *avoir*, "to have." The closest equivalent is *ginyen*, meaning to gain or earn. So in Thomonde the people take nothing for granted; one doesn't "have" good health, one earns it. A trip to Thomonde will quickly shatter stereotypes about the poor—that the poor are lazy or dumb or lack family values. The Thomondois as a group are industrious and eager for knowledge about their health. Their entire lives are family centered—materially, in the sense that life evolves around the inherited family plot of land; but also spiritually—children, old folks, and ancestors are all revered. They're also community centered. There's not only a knowledge of and concern for all of one's neighbors but also a communal way of getting things done, whether it's building a house or clearing a field. These are traditions that started in the days of slavery. In this milieu it has been much easier to build a health care system than one might think.

To Medishare's credit, we haven't just thought about health—we've supported education, nutrition, and, most importantly, created meaningful jobs, perhaps the most effective way to fight the infectious diseases that afflict the poor. The uplifting of socioeconomic status that Medishare has brought to Thomonde is the second secret of our success. We've created a rural middle class in the health care sector. Our community health care workers cost us about $100 per month—a pittance by U.S. standards, but a decent salary in a country with an average per capita income of $200 per year. These funds have percolated throughout the commune, contributing to the well-being and elevating the standard of living and quality of life of even the poorest Thomonde peasant.

Medishare is using health as a fulcrum to leverage community development. Ellen, Medishare's executive director, has organized the commune and provided the resources to produce a protein-enriched powder that, after just adding water and boiling, becomes a nutritious, culturally accepted porridge. Called Akamil, this porridge will not only treat the malnourished children of Thomonde but will also be sold for a modest profit in other communities in Haiti.

Impressed with the progress Medishare has initiated in Thomonde, another foundation, Fonkoze, has started a program in micro-loans, to further reduce the level of poverty. Totally unexpected developments are happening spontaneously, thanks to Medishare's investment in the people of Thomonde. Nimi, our cook, is one example. She saved the money we pay her to cook for the Medishare teams and opened a school for cooking and baking. Evidently, there's a market for this, particularly the baking skills. Cakes are needed to celebrate all major life events, and it's no small feat to be able to bake a cake over an open charcoal stove. Nimi has now graduated two classes of chefs, who study with her for a full year and receive a certificate upon graduation. In addition to sharing her culinary secrets, Nimi teaches nutrition and food hygiene, knowledge and skills she learned from Medishare. These are critical advances for public health in rural Haiti.

As Medishare celebrates its 10th anniversary, it has far exceeded my original expectations. Every day now, starting at 6:00 a.m., Nimi and her crew brew several pots of Haitian coffee, scramble eggs, and cook Haitian spaghetti to send off our 50 community health workers on their daily rounds. The workers pack their coolers with ice, so their vaccines won't spoil, and stuff their satchels with Direct Observed Therapy forms and precious medicines. By 7:00 a.m. they set off—some on foot, some by horse, and some by motorcycle—to the farthest reaches of the commune, full of zeal and enthusiasm for their work and a loving concern for their patients that I rarely see anymore in America.

In addition to treating AIDS and tuberculosis, our workers have launched an immunization campaign. In 2002, Thomonde had the worst immunization rates in Haiti. Now, 98 percent of the children

have been immunized. In 2005 the Pan American Health Organization plans to finally begin a campaign to eradicate rubella from Haiti.
Thomonde will be one of the few rural communities with an infrastructure to implement the project. By 7:30 a.m. the school children
head for class, replete with book bags and color-coded uniforms. If
they show signs of malnutrition, they'll receive Akamil at school, in
exchange for their parents keeping them enrolled. Pregnant women
and patients on Direct Observed Therapy also get Akamil, flavored
with cinnamon and vanilla. The dispensary opens at 9:00. On the
average, 200 patients pass through its doors each day.

Meanwhile, before dawn, while the lycee students up in Cap
Haitien are pacing and reciting their *devwa* ("homework"), our family medicine residents are reviewing their admissions from the night
before in anticipation of attending rounds. Patients line up each day
at the family practice center for immunization and family planning.
The program is now sustained through funding from the U.S.
Agency for International Development. Haiti's ministry of health is
encouraging us to expand training of family doctors, and we're hoping to open a second program based in Thomonde as soon as we've
completed construction of a new hospital and clinic. Blessed with
knowledgeable and enthusiastic Haitian partners, day by day, we're
making a difference.

Thanks to the visit by the president and the dean, the issues
between Medishare and the University of Miami are now history.
The dean will be sending a team of eye doctors and students to give
eye care to the people of Cange and Thomonde on a regular basis.
President Shalala is helping us connect with some major international donors. With their help, and the help of a prominent Miami
family, the Jay W. Weiss Center for Social Medicine and Health Inequality has been founded at the University of Miami-Miller School
of Medicine. Almost 30 faculty members have signed on, with a
variety of projects planned at home and abroad to help those in
need. Our work in Haiti serves as the center's flagship, leading the
way. Working with Paul Farmer, the center will train a special group

of residents in family medicine, pediatrics, and internal medicine in the skills necessary to be effective in international health—teaching them not just how to treat the diseases associated with poverty but more importantly how to attack its root causes.

More than 100 first-year medical students—that's two-thirds of the class—have signed up to volunteer for Project Medishare. They are busily planning, along with our 50 second-year students, this year's health fairs. Over the years more than 500 medical students, doctors, and nurses have contributed to Haiti's health under Medishare's umbrella. Other medical schools—notably, George Washington and Northwestern—are partnering with us to expand services in the communities around Thomonde.

Medishare's collaboration with Paul's charity *Zanmi Lasante* (Partners in Health) has allowed us to expand our community health workers and created an environment in which the Haitian doctors and nurses we've hired for Thomonde have the backup and support they need. In turn, Paul has hired one of the graduates of our family medicine training program to work in Cange. "Send me more," he says, when I ask him how our doctor is doing. "He can do it all!" To date, we've graduated 12 doctors from our residency program. All but two are now working as family doctors in rural Haiti. We'll be producing at least five Haitian family doctors per year and will hopefully increase that number to 10 in the near future. My still-unfulfilled dream is to start more training programs and link them to the training of nurses and community health workers. Thomonde, with its seasoned and skilled health care workers, with Marie (Medishare's nurse), and Delva's spacious new guest house would be an ideal training site, assuming we can raise the funds to build a new hospital and clinic.

Medishare owes a great deal to Dr. Paul Farmer. The example he sets in terms of sacrifice, dedication, cultural competency, and solidarity with the people, plus his commitment to health as an issue of social justice and his unwillingness to accept a double standard of care for poor people is remarkable. Medishare, however, is now making its own contributions to the evolving strategies to improve health care for the poor. Paul, in his writings, eloquently makes the point that effective treatment with unequal access to care is an injustice. Through our health fairs, patient education programs, immunization campaigns, and training program, we've extended Paul's concept to include not just treatment but also screening, prevention, and primary care. Second, we steadfastly believe that, ultimately, through education, the torch must be passed to Haitian providers.

During the past 10 years, my students and I have made over 100 trips to Haiti. Our project has expanded, and it is still growing today. Yes, we've experienced our share of scorching heat, torrential rains, political turmoil, flat tires, and mud holes. We've also had total strangers help us change those tires, pull us out of those mud holes, and shelter us from the tribulations that are part of daily life in one of the poorest countries in the world. In the early days, I frequently traveled alone. Now, I never lack for friends and companions. Invariably, the Haitian people have given much more to me—lessons of courage, patience, ingenuity, and mysticism—and countless opportunities to make a difference. More than unlocking the mysteries of Voodoo and the zombies, more than making my students and me better doctors, my experiences have shown me how to live a better life. *"Ayiti te mete yon wanga sou mwen!"* ("Haiti cast a spell on me!").

Postscript

Suz: Hi, Dad.

Dad: Hi, Suz. How's it going?

Suz: Winter in Ann Arbor. Ahrgghhh!

Dad: This too shall pass. I was really proud of the work you did during our winter break trip.

Suz: Yeah, that was fun.

Dad: I mean, it was the first time anyone had ever done dental screenings in the remote corners of the commune. And then to diagnose congenital syphilis by the shape of that child's teeth! Wow!

Suz: Yeah, my classmates couldn't believe it!

Dad: Listen, a funny thing happened to the dental unit we shipped down for Thomonde.

Suz: Oh?

Dad: We sent it down with a mammogram machine for a cancer support group in Petionville. The dental unit went to Petionville, and the mammography machine to Thomonde!

Suz: Oh, no!

Dad: Not to worry. We'll have it all sorted out by the next time you return.

Suz: I'm so excited!

Dad: Me, too! Dental care for Thomonde! Who ever would have thought it!

Suz: I know. I've got faculty and students signed up to go, and 500 toothbrushes. . . .

Dad: A little bit of prevention will go a long way. Hey, Suz— would you look up a book in the library?

Suz: Dad you're getting so old! Nobody goes to the library anymore! Just download whatever you're looking for off the Internet.

Dad: No Suz. This won't be on the Internet. It's a book written in Creole probably about 30 years ago. *Fundamentals of Dental Care*, author Régis _____. It may not even be in the library, but I'd just like to know.

Suz: Wow, that's cool. If it's in Creole, we could really use it! How'd you find out about it?

Dad: It's a long story.

Suz: Dad, you're getting a little cryptic on me.

Dad: You know what a slow typist I am! I'll tell you the whole story in Haiti.

Suz: You're testing my patience!

Dad: Hey, you remember the kid who was dumped in your arms at Bercail Bon Berger on your second trip?

Suz: Dad! Of course, I do!

Dad: I saw him during my last trip. He's doing great! A real *gwo-neg*! Lots of personality!

Suz: That's great. I love you!

Dad: I love you, too!

Suz: Bye!

Dad: Bye!

A Poem for Haiti*
By *Maya Angelou*

Homo sum. Humani nil a me alienum puto!

I am a human being
Nothing human can be alien to me!
This statement was made
By Terentius Afer.
Terentius was a slave,
Sold to a Roman senator
Freed by that senator.
He became
The most popular playwright in Rome.
Five of his plays
And that statement has
Come down to us from 154 B.C.
That man
Not born white or free
Or with any chance
Of becoming a citizen
In the Rome of his day,
He said, "I am a human being.
Nothing human can be alien to me!"
I have ingested and digested that statement.
It activates me
And motivates me to say,
I'm yours
Haiti
Yours . . ."

---

*Composed extemporaneously by Dr. Maya Angelou, after a screening of "Once There Was a Country," a documentary detailing the work of Medishare and Partners in Health in Haiti. The screening was held at the Open Society Institute, New York, July 7, 2005.

Index